Intellectual and Imaginative Cartographies in Early Modern England

Taking as its focus an age of transformational development in cartographic history, namely the two centuries between Columbus's arrival in the New World and the emergence of the Scientific Revolution, this study examines how maps were employed as physical and symbolic objects by thinkers, writers and artists. It surveys how early modern people used the map as an object, whether for enjoyment or political campaigning, colonial invasion or teaching in the classroom. Exploring a wide range of literature, from educational manifestoes to the plays of Marlowe and Shakespeare, it suggests that the early modern map was as diverse and various as the rich culture from which it emerged and was imbued with a whole range of political, social, literary and personal impulses.

Intellectual and Imaginative Cartographies in Early Modern England, 1550–1700 will appeal to all those interested in the History of Cartography.

Patrick J. Murray, PhD, is a researcher specialising in early modern literature. His primary research and teaching interests focus on the interdisciplinary interfaces of cartography, literary representation and cultural fashioning in the period 1550–1750.

Routledge Studies in Renaissance and Early Modern Worlds of Knowledge

Series Editors:

Harald E. Braun (University of Liverpool, UK) and Emily Michelson (University of St Andrews, UK)

SRS Board Members:

Erik DeBom (KU Leuven, Belgium), Mordechai Feingold (California Institute of Technology, USA), Andrew Hadfield (Sussex), Peter Mack (University of Warwick, UK), Jennifer Richards (University of Newcastle, UK), Stefania Tutino (UCLA, USA), Richard Wistreich (Royal College of Music, UK)

This series explores Renaissance and Early Modern Worlds of Knowledge (c.1400–c.1700) in Europe, the Americas, Asia and Africa. The volumes published in this series study the individuals, communities and networks involved in making and communicating knowledge during the first age of globalization. Authors investigate the perceptions, practices and modes of behaviour which shaped Renaissance and Early Modern intellectual endeavour and examine the ways in which they reverberated in the political, cultural, social and economic sphere.

The series is interdisciplinary, comparative and global in its outlook. We welcome submissions from new as well as existing fields of Renaissance Studies, including the history of literature (including neo-Latin, European and non-European languages), science and medicine, religion, architecture, environmental and economic history, the history of the book, art history, intellectual history and the history of music. We are particularly interested in proposals that straddle disciplines and are innovative in terms of approach and methodology.

The series includes monographs, shorter works and edited collections of essays. The Society for Renaissance Studies (http://www.rensoc.org.uk) provides an expert editorial board, mentoring, extensive editing and support for contributors to the series, ensuring high standards of peer-reviewed scholarship. We welcome proposals from early career researchers as well as more established colleagues.

For more information about this series, please visit: https://www. routledge.com/Routledge-Studies-in-Renaissance-and-Early-Modern-Worlds-of-Knowledge/book-series/ASHSER4043

Intellectual and Imaginative Cartographies in Early Modern England

Patrick J. Murray

Routledge
Taylor & Francis Group

LONDON AND NEW YORK

First published 2023
by Routledge
4 Park Square, Milton Park, Abingdon, Oxon OX14 4RN

and by Routledge
605 Third Avenue, New York, NY 10158

Routledge is an imprint of the Taylor & Francis Group, an informa business

© 2023 Patrick J. Murray

British Library Cataloguing-in-Publication Data
A catalogue record for this book is available from the British Library

Library of Congress Cataloging-in-Publication Data
A catalog record has been requested for this book

ISBN: 978-1-032-06025-5 (hbk)
ISBN: 978-1-032-06026-2 (pbk)
ISBN: 978-1-003-20037-6 (ebk)

DOI: 10.4324/9781003200376

Typeset in Times New Roman
by KnowledgeWorks Global Ltd.

Contents

Acknowledgements

Francis Bacon observes that learning is comprised of three separate activities: reading, writing and conference. In the process of composing this book, the sagacity of this observation has been increasingly apparent. Having read and written, I am mindful of the vital role of conference in the shaping of the intellectual faculties. As such, I would like to extend my sincere gratitude to a number of people for their assistance in the long gestation of this project.

This includes my fellow scholars at the University of Glasgow, who have provided a nourishing intellectual environment for a scholar of early modern studies during the completion of the doctoral thesis precipitating this book. I would also like to express my appreciation to the various members of the University of Strathclyde's Renaissance Reading Group, who have graciously taken the time to consider parts of this thesis and provided valuable feedback. This includes (but is not limited to) Alison Thorne, Jonathan Hope, Steven Veerapen, Erica Fudge, Peter Argondizza, Heather Froehlich and Elspeth Jajdelska. I am also thankful to Jonathan Sawday, Jerry Brotton, Sarah Bendall, Katherine Heavey, Donald Mackenzie and Annaleigh Margey for their helpful advice along the way. Moreover, the responses of the peer reviewers in the publication process, as well as those of my series editors, Emily Michelson and Harald Braun and Michael Greenwood at Routledge, have been invaluable.

I would also like to thank The Carnegie Trust for the Universities of Scotland for their inestimable support. In particular, I would like to express my thanks to Barbara Bianchi and Julie Black, whose ready assistance on all matters, great and trifling, has been greatly appreciated.

I am especially grateful to Willy Maley and Rob Maslen. Both have proved invaluable guides during the long, occasionally onerous but ultimately rewarding task of doctoral research. I would like to thank both for imparting their wisdom. I hope this study does their tutelage justice.

Finally, I would like to acknowledge the support of my mother and father during the initial composition period. Furthermore, I am immensely grateful for the encouragement of Heather May, without whom this book would not have reached publication.

It is dedicated to her.

Introduction
Weaving the Net

Composed in 1611, John Donne's 'The First Anniversary: An Anatomy of the World' presents an elegiac rumination on the life and death of Elizabeth Drury, daughter of his close friend and sometime patron Robert Drury. Reflecting upon the vagaries of humanity's engagement with the corporeal world, Donne addresses what he views as its inconsistency of order and disorder:

> We think the heavens enjoy their spherical,
> Their round proportion, embracing all;
> But yet their various and perplexed course,
> Observed in divers ages, doth enforce
> Men to find out so many eccentric parts,
> Such diverse downright lines, such overthwarts,
> As disproportion that pure form; it tears
> The firmament in eight-and-forty shares,
> And in these constellations then arise
> New stars, and old do vanish from our eyes;
> As though heaven suffered earthquakes, peace or war,
> When new towers rise, and old demolish'd are.
>
> (251–262)[1]

The target of Donne's critique is the human presumption of uniformity across the spheres: 'We think the heavens enjoy their spherical, | Their round proportion, embracing all'. Exploring upon the apparent incompatibility between systematised man and organic nature, Donne describes humanity's striving for order as 'eccentric', 'diverse downright lines' and 'overthwarts'. In response to the 'disproportion' of the natural and its 'varied and perplexed course', we vainly struggle for discipline and regularity. Such desires manifest an attempted imposing of order: while '[n]ew stars, and old do vanish from our eyes', there persists an attempt to project regulated structure onto formlessness and apparent chaos. However, in doing so, 'Men', Donne's speaker observes:

DOI: 10.4324/9781003200376-1

> [...] have impaled within a zodiac
> The free-born sun, and keep twelve signs awake
> To watch his steps; the Goat and Crab control,
> And fright him back, who else to either pole,
> Did not these tropics fetter him, might run.
>
>   (263–266)

The ferocity of this imagery – impaling the sun and 'fright[ing]' it to the extent that, were it not for the bounds of the 'tropics', it would disappear – is striking. Characterising the introduction of an ordered system of knowledge (in this case, the 12 signs of the astrological zodiac) into the universe as an unnatural action against the heavens, it locates the schematising impulse of cosmographical surveying within a discourse of violence. Donne continues his attack on systematisation, articulating with characteristic elan an even more remarkable conceit:

> For of meridians and parallels
> Man hath weaved out a net, and this net thrown
> Upon the heavens, and now they are his own.
> Loth to go up the hill, or labour thus
> To go to heaven, we make heaven come to us.
> We spur, we rein the stars, and in their race
> They're diversely content t'obey our pace.
>
>   (278–284)

In these lines, which call to mind the gridded nature of cartography, mapping is figured as a means for humanity to exercise control, order and regulation on the world. Thus, the map is evoked, its 'meridians and parallels' characterised as a 'net' to capture the object.

More emotively, there is an attendant declaration of mapping's predisposition towards appropriation and ownership. Man has 'weaved' the net, 'thrown' it 'upon the heavens' and attempts to 'spur' and rein' the stars in 'their race', believing their motions to parallel events on earth. The imperative of possession in cartographic practice is underlined by the trope of the net, connoting capture and subjugation. Humanity, in Donne's view, seeks to align the terrestrial with the heavenly, attempting to formulate a rational model of the cosmos. Superficially disordered, the motions of the planets will be made, and here the possessive pronoun is notable, 't'obey our pace'.

Donne's emphasis is primarily upon the drawing of star maps, used for centuries to record observations about the motions of the heavens, and through the use of astrological theory to prognosticate events to come.[2] Yet Donne also draws in terrestrial cartography to his appraisal of the belief in the power of reason and order in the physical sphere. Earlier in the poem, the speaker alludes to the work of Claudius Ptolemy, referring to the Greco-Roman polymath's highly influential projection system, which conceptualised the heavens in 48 separate sections, remarking 'It tears | The firmament in eight and forty shares' (267–268). Any attempt at ordering is intrinsically vexed, especially when it attempts to capture the nature of the earth:

> But keeps the earth her round proportion still?
> Doth not a Tenerife or higher hill
> Rise so high like a rock, that one might think
> The floating moon would shipwreck there and sink?
> Seas are so deep that whales, being struck today,
> Perchance tomorrow scarce at middle way
> Of their wish'd journey's end, the bottom, die.
> And men, to sound depths, so much line untie
> As one might justly think that there would rise
> At end thereof one of th'antipodes.
>
> (285–294)

Donne queries the plausibility of the subjective measurement and recording of the earthly. As the globe lacks all regularised 'proportion', containing hills so high that they could shipwreck the moon and seas so deep that they drown whales, so its geography eludes any sense of scientific regularity and proportionality.

Such hyperbolic claims regarding the earth's geography see the poet utilising the rhetorical device of auxesis – the intentional overstating of a particular fact – as a means to interrogate a paradox: how can man map with ordered 'meridians and parallels' encompass the essentially disordered? Jonathan Sawday provides an eloquent analysis of the dichotomy at the heart of Donne's poem, and in particular, the map/net image. 'For Donne', Sawday writes, 'the net suggested an unreservedly modern conception of knowledge, but it also expressed human vanity, rather than human achievement'. 'The heavenly net', Sawday continues:

> is woven out of the new philosophy of reason, conceived of as "meridians and parallels", the celestial longitude and latitude of the astronomy of Copernicus and his contemporaries. Donne's net, in other words, functions in both a positive and negative sense. It is positive in that human reason *appears* to be making sense, at last, of the eccentric orbits of the heavenly spheres, which are now subjected to a more precise mathematics. But even as this advance is celebrated, the negative image of the net glances back to the mythological nets of old, woven from cunning rather than reason, which are designed to capture and subdue mysteries which should remain beyond reason.[3]

Sawday's reading contextualises Donne's net within the febrile environment of early modern technological development. For him, the map 'The

Anniversary', with its 'meridians and parallels', serves as a wider allusion to a broader schema of failed knowledge when faced with the irresolvable problem that 'none ends where he begun' (842). While Alma B. Altizer suggests that Donne criticises humanity's false presumptuousness in creating a correspondence between the heavenly and the human,[4] there is also a more profound element to the rhetorical flourish – the foregrounding of the role of the mapper, the producer and wielder of the net. 'Even in Donne's time', Dennis M. Lensing argues, 'the cartographic projects associated with astronomy are connected with appropriation and ownership'.[5] Donne explicitly evokes this sense of appropriation and ownership with his paralleling of maps and manufactured nets. Pointedly, Donne articulates not just the product but also the process of production and the producer: 'man hath weaved out a net'.

Production lies at the heart of Donne's reticulated model, and the act of production necessitates a producer. In foregrounding cartography and its systematising mode of representation – the gridded 'meridians and parallels' on which the image turns – within a critique of scientific rationalisation, the speaker emphasises the fundamental contradiction at the heart of mapping's 'neutral' enterprise. The striving for an objective depiction of geography is by its very nature, subjective. As the historians of science Lorraine Daston and Peter Galison remind us, 'objectivity is a multifarious, mutable thing, capable of new meanings and symbols'.[6] As a result of this multifariousness and mutability, maps become more than a representation of a selected geography. The portrayal and projection of reality and real landscapes is possible only through a medium which is in itself manufactured and mitigated. By the necessary presence of a mapper, the map is infused with a range of personal, economic and socio-political imperatives.

The understanding of the map as a socially engineered artefact has been a hallmark of cartographic theory since the foundational work of J.B. Harley. Drawing on a number of theoretic paradigms from discrete fields such as psychoanalysis, sociology, deconstruction, economic history and postcolonial theory, Harley has helped establish the map as a product of a complex system of interacting discourses, including governmental power, economic hegemony and expressions of nationalism. The primary benefit of Harley's work has been the broadening of the analytical tools available to historians of cartography.[7] As J.H. Andrews playfully notes:

> The peculiar quality of Harley's achievement may best be appreciated by inventing a game for map historians: Tabulate the specialized vocabularies of cartography, philosophy, and the aggregate of all other academic subjects in three columns. Choose one term at random from each column. Whoever combines these terms into the most arresting and plausible short sentence wins the round.

Various examples illustrate the interdisciplinarity of Harley's work. In 'Silences and Secrecy: The Hidden Agenda of Cartography in Early Modern

Europe',[8] Harley employs the French philosopher Michel Foucault's theories of social and political power to show how in sixteenth- and seventeenth-century Europe '[t]he map image itself was becoming increasingly subject to concealment, censorship, sometimes to abstraction or falsification'.[9] In 'Power and Legitimation in the English Geographical Atlases of the Eighteenth Century', the author utilises historian Joseph Rouse's study *Knowledge and Power: Toward a Political Philosophy of Science* and its ideas about standardisation in science and scientific knowledge to interrogate how maps project and legitimise standardised images of control.[10] And in his essay 'New England Cartography and the Native Americans',[11] Harley draws on theories regarding the growing prominence of Anglicisation in place names to investigate how maps of discovered lands produced by the discoverer evince underlying colonial bias.

Probing 'mapping moments' in a wide and varied range of literature, this book examines how early modern thinkers and writers engaged with the practice of world describing on an intellectual and imaginative level. It aims to investigate what I have termed cartography's 'multidisciplinary inclination', whereby mappers such as John Speed sought to draw on varied intellectual pursuits such as historiography, geology and literary composition to immerse cartographic production in a wider, cross-pollinating intellectual landscape. It throws fresh light on key texts of scholarly investigation of the inter-relationships between cartography and power in the early modern period, for Christopher Marlowe's *Tamburlaine*, William Shakespeare's *King Lear* and Michael Drayton's poem *Poly Olbion*. In addition, it broadens the scope of recent cartographic scholarship to consider more obliquely map-inflected material, such as Wenceslaus Hollar's Civil War portraits of battlefield commanders, John Donne's metaphorical maps and the cartographical musings of Francis Bacon. In doing so, it examines the motivations, influences and aims behind the Renaissance mapper's hand; explores the role of cartographic objects in the educational environment; and charts how the map figured as an object of the imagination, a catalyst for mental as opposed to physical travel.

Motivated by the Harleian notion of the map as more than a verisimilar copy of a tract of land primarily, this study also pays attention to Christian Jacob's advocacy of a 'cultural contextualisation' of cartographic practices.[12] Jacob writes:

> There are yet other ways of looking at maps. The historian of cartography can consider maps in isolation, as self-defined artefacts to be classified and analysed. Or an attempt can be made to understand maps within the culture that produced and used them, so long as such a contextual approach does not lose sight of the map itself.[13]

Incorporating a more comprehensive intellectual scope beyond a purely material consideration of the map, I explore how maps were conceptualised, produced, studied and used by early modern English and British culture.

Paying heed to Jacob's suggestion that 'there are yet other ways of look-ing at maps', I consider the cartographic document as a product of a wider cultural discourse in a unique period in the history and development of car-tographic practice.

The two centuries of print revolution and colonial expansion between 1500 and 1700 witnessed an exponential increase in the sophistication, exactitude and proliferation of mapping in Europe. Cartographers such as Battista Agnese, Abraham Ortelius, and Gerard Mercator, aided by the refinement of instruments like the astrolabe and the quadrant and enlightened by the discoveries of 'new' lands, produced maps and atlases unprecedented in scope and accuracy. In the Netherlands, the flourishing of commercial cartography saw the emergence of the great map publishing dynasties of Hondius and Blaeu. As the many reproductions of maps and globes in the paintings of the putative Dutch Golden Age demonstrate, cartographic objects became a feature of everyday life, decorating the households of middle-class and mercantile families in Leiden, Haarlem and Amsterdam.[14]

Such developments infiltrated a vast array of social, civic and political spheres. '[E]arly modern maps and mapping practices', writes Richard Helgerson, 'had their part in national consolidation, overseas expansion, humanist and Reformation historicism, emerging agrarian capitalism, scientific revolution, and a general abstracting of time and space'.[15] In the febrile cultural atmosphere of the Atlantic archipelago in the sixteenth and seventeenth centuries, the visibility of maps grew significantly. The natural philosopher, royal doctor and sometime-spirit conjuror John Dee provides an insight into the extent of the enthusiasm for maps in late sixteenth-century English intellectual culture:

> While, some, to beautifie their Halls, Parlers, Chambers, Galeries, Studies, or Libraries with: other some for thinges past, as battels fought, earthquakes, heauenly fyringes, & such occurentes, in histories men-tioned: therby liuely, as it were, to vewe the place, the region adioyning, the distance from vs: and such other circumstances. Some other, pres-ently to vewe the large dominion of the Turke: the wide Empire of the Moschouite: and the litle morsell of ground, where Christendome (by profession) is certainly knowen. Litle, I say, in respecte of the rest. &c. Some, either for their owne iorneyes directing into farre landes: or to vnderstand of other mens trauaile. To conclude, some, for one purpose: and some, for another, liketh, loueth, getteth, and vse Mappes, Chartes, & Geographicall Globes. Of whose vse, to speake sufficiently, would require a booke peculiar.[16]

In a similar fashion, the educational theorist Thomas Blundeville begins his tract *A Briefe Description of the Universal Mappes and Cardes, and of their Uses* (1589) with the remark 'I Daylie see many that delight to looke on Mappes'.[17,18]

The observations of Blundeville and Dee give an indication as to the extent of cartographic consciousness in the period. Dee's comment, in particular, demonstrates the visibility of maps. Moreover, it also draws attention to the manifold purposes to which they were deployed. Some use maps to 'beautifie their Halls, Parlers, Chambers, Galeries, Studies, or Libraries', revealing a prevailing appreciation of the aesthetics of cartography. Some use them as educational aids to elaborate upon 'histories mentioned', gesturing towards cartography's role in a broader system of knowledge. Some use them for the mathematical measuring of distance. Others still use them as means to 'thereby liuely [...] vewe the place', an intriguing form of mind travel that, many Renaissance thinkers believed, was facilitated by the presence of the map. Indeed, such is the variety of uses for the map, Dee concedes, that 'to speake sufficiently [of them], would require a booke peculiar'.

Dee's observations remind us of the versatile possibilities the map offered the map-reader, from decorative to educational, geographical to historical and pedagogical to leisurely. Previously largely derivative of European sources, English and British cartography, nurtured by a collective growing sophistication in the educational approach to mapmaking and enhanced by the work of individual mapmakers such as Emery Molyneux, John Norden and Aaron Rathborne, enjoyed an elevation in cultural status. 'In the late sixteenth century', notes Donald K. Smith, 'the new introduction of cartographic representation to a widespread, literate public, brought about a shift in the way terrestrial space could be represented'. Manifesting a 'whole new way of looking at the world', this technological development would have a profound effect on conceptualisations of space and its attendant ideas such as community, identity and nationality:

> Improvements in surveying technology made it possible for increasing numbers of individuals to measure and record the shape and order of the country, and with the development of scale maps in the early 1540s a sense of confidence in the precision and accuracy of spatial representation became possible.[19]

As the emergent British Empire developed, the British Isles and the British nation were conceptualised cartographically in the surveys of Christopher Saxton. Saxton's popular *Atlas of the Counties of England and Wales* (1579) was the first volume of its kind in England and fuelled a burgeoning interest in descriptions of the landscape. Subsequent cartographers would attest to Saxton's influence: John Norden, in an address 'To all Courteous Gentlemen, Inspectators and practitioners in Geographie', which prefaces his *Preparatiue to his Speculum Britanniae* (1596), pointedly pays tribute to the 'most painful & praiseworthie labours' of his predecessor.[20]

Alongside Saxton's cartographies, writers such as William Camden and Michael Drayton would also produce textual descriptions of the landscape, signposting the ways in which the passion for world-describing engendered

Ditchley Portrait of Elizabeth 1st by Marcus Gheeraerts the Younger oil on canvas, circa 1592. (Reproduced by permission of The National Portrait Gallery, London)

into new and exciting textual forms. Chorographies, or the study of specific localised areas characterised by a descriptive tone and an articulation of local geographic, demographic and cultural features, began to flourish. In his influential treatise *The Cosmographical Glasse* (1559), William Cuningham sets out how written descriptions of the world emerged, coexisted and prospered alongside the typically imagistic medium of cartography

> For lyke as Cosmographie describeth the worlde, Geographie th'earth: in lyke sorte, Chorographie, sheweth the partes of th'earth, diuided in themselues. And seuerally describeth, the portes, Riuers Hauens, Fluddes, Hilles, Mountaynes, Cities, Villages, Buildinges, Fortresses, Walles, yea and euery particuler thing, in that parte conteined.[21]

Michael Drayton's *Poly-Olbion*, a poem of nearly 15,000 lines, is emblematic of this genre: moving from county to county, it describes the various natural landscapes of sections of England and Wales. In a similar fashion, though on a smaller scale, William Lambarde's *A Perambulation of Kent* (1570) sought to depict the 'description, historye and customes of that shire'.[22] Such was the popularity of this type of literature that works which could be defined as generically geographic increased exponentially. John M. Adrian estimates that texts which he defines as 'focus[sing] on or organised around geographical units' increased more than 20-fold between the 1520s and the 1590s.[23]

   If the proliferation of these texts indicates the growing popularity of geographical perspectives, the multi-disciplinary characteristics of such works testify to their intellectual brio. The combination of geographic detail, local history, demography, and economic and sociological description by the likes of Lambarde hints at the fluid character of early modern geographical texts and their multi-disciplinary scope. Illustrative of Renaissance geography's dynamic epistemology is Camden's opus, which included in later publications of his chorography of Britain maps by Saxton and Norden. In a 1610 edition of *Britannia* translated by Philemon Holland, the preface reads:

> Many have found a defect in this worke that Mappes were not adjoyned, which doe allure the eies by pleasant portraiture, and are the best directions in Geographicall studies, especially when the light of learning is adjoyned to the speechlesse delineations. Yet my abilitie could not compasse it, which by the meanes and cost of George Bishop, and John Norton is now performed out of the labours of Christopher Saxton, and John Norden, most skilfull Chorographers.[24]

In an echo of Dee and Blundeville's reflections on the visual gratifications of cartography, Camden exhibits an awareness of the pleasure of maps 'which doe allure the eies by pleasant portraiture'. While the original 1586 publication was solely textual, later editions are compelled by the growing cartographic enthusiasm to incorporate 'Mappes' into the revised chorography. It recognises the prevailing belief that 'the best directions in Geographicall studies [occurs] especially when the light of learning is adjoyned to the speechlesse delineations'. In redressing the lack of maps in previous editions, the preface attests to the flourishing appetite for cartography among seventeenth-century readers.

The later edition of *Britannia* also gestures towards the diversity of early modern classifications of geographical science: Saxton and Norden, putatively mappers on both a local and national scale, are identified as chorographers. They are aligned to figures such as Lambarde, whose wholly textual *Perambulation* shares a similar purpose – the delineation of geographical space – if not an entirely similar medium. John Speed's *Theatre of the Empire of Great Britaine* (1611) meanwhile exemplifies the ways in which early modern geographic literature took on an increasingly multimedia character to satisfy the demands of map enthusiasts. Representing one of the most ambitious cartographic projects of the age in its scope and physical scale, it contains a single visual survey for each county of England and Wales, and also a map of Scotland and each of the four Irish provinces. Each section incorporates both cartography and chorography, with a map presented alongside an extensive textual narrative encompassing various local facts. Moreover, Speed's work demonstrates that the focus on particular, localised areas coexisted alongside a more expanded form of geographical description.

Speed's *Theatre* also reflects the lofty ambitions of the practice of world-describing during the Renaissance. Deliberately adopting and adapting the title of Ortelius's *Theatrum Orbis Terrarum* (literally translated as 'Theatre of the World') (1570), this folio volume of over 550 pages had, as the inclusion of 'Empire' in its title suggests, a grandiose scope. Infused with an almost evangelical zeal, the *Theatre*'s sense of divine ordination in its cartographic construction of Great Britain is evident in its preamble. In the preface to the reader, Speed describes the 'building' of the *Theatre*:

> And applying my selfe wholely to the frame of this most goodly Building, haue as a poore labourer carried the carued stones and polished Pillars from the hands of the more skilfull Architects to be set in their fit places, which here I offer vpon the altar of loue to my country, and wherein I haue held it no sacrilege to rob others of their richest iewels to adorn this my most beautifull Nurse, whose wombe was my conception, whose breasts were my nourishment, whose bosome my cradle, and lap (I doubt not) shall be my bed of sweete rest, till Christ by his trumpet raise me thence.[25]

The overt religiosity of Speed's proclamation – formulated in phrases reminiscent of the Book of Isaiah's figuring of God as a builder[26] and concluded by a reference to Judgement Day when 'Christ by his trumpet raise me thence' – imbues the atlas with an evangelical zeal from the outset.

The figuring of 'Britaine' as a 'most beautifull Nurse', a 'wombe' where the author has been nourished conveys the patriotic sentiment at the heart of Speed's cartographic project, one bound up with a highly formalised conception of the British state. Speed's atlas, as J.H. Andrews suggests, 'carr[ies] a very powerful political message'[27] affecting a range of sociocultural assumptions regarding British identity, culture and history. For example, in the inclusion of all four nations comprising the Atlantic archipelago, the mapper displays a singular identity denuded of its internal tensions and projecting a settled hierarchy: 'Whatever his attitude toward empires and empire-building', writes Andrews, 'Speed saw no reason why Ireland should be bracketed with Virginia or Bermuda. Instead he put it in the first British atlas of the British Isles, ranking in prominence above Scotland, although well below England and Wales'.[28] In Speed's atlas, geographical science becomes a tool of British nationalist and unionist expression.

Furthermore, Speed, as his introduction with its conception of Britain as a 'Nurse' and 'wombe' demonstrates, also makes use of personified representations of landscape. Such a device was a recurrent motif in contemporary cartographic and geographic texts and stemmed from the desire to project the land as a living and well-organised entity with a place for everything and everything in its place. Mappers and chorographers utilised the technique of personification of the landscape to express a sense of ordered structure. In doing so, they borrowed heavily from another evolving discipline of the age: anatomy. As my analysis of Speed and Drayton, in particular, will show, this deliberate adoption of another disciplinary approach was most evident in cartography's use of the lexicography of another seemingly discrete discipline: anatomy. I pay particular attention to the relationship between figurative imaginings of cartography as anatomy and the land as body. In an important study of the links between cartography and anatomy in early modern England, Valerie Traub has drawn attention to the 'subjection of the bodily interior to anatomy's scopic, penetrative, reifying gaze'.[29] Anatomy's reification process, transposed onto the land, consequently sought to figure geographical landscape as a direct analogue of human biology. Rivers were conceived as veins, provinces as limbs and cities as vital organs, all functioning together to form the perfect nation state or, in the words of Speed, the 'Eden of Europe'.[30]

Symbolically, this body was frequently gendered. The *Theatre*'s presentation of Britain as a 'wombe' evokes a feminised image of the landscape and this technique was reciprocated in other contemporaneous geographical texts. In Drayton's textual chorography *Poly-Olbion*, the land assumed a female form through the use of feminised personification and also the striking quasi-cartographic female anatomy-cum-map of Britain, which adorns the frontispiece. Drawing the geographically focused chorography into the broader issue of gender politics, the female presence in Drayton's poem foregrounds the complex set of impulses behind anthropomorphic representations of geography. As Annette Kolodny has observed in early descriptions of the New World, and in particular North America, in her cleverly titled study *The Lay of the Land: Metaphor as Experience and History in American Life and Letters*, feminine figurations of landscape in the fifteenth, sixteenth and seventeenth centuries carried with them a pernicious undertone of conquest and rape:

> [E]ver since the first explorers declared themselves virtually 'ravisht with the ... pleasant land' and described the new continent as a 'Paradise with all her Virgin Beauties' [...] [t]he human, and decidedly feminine, impact of the landscape became a staple of early promotional tracts, inviting prospective settlers to inhabit 'valleyes and plaines streaming with sweete Springs, like veynes in a natural bodie,' and to explore 'hills and mountains making a sensible proffer of hidden treasure, neuer yet searched'.[31]

The framing of the land as feminine and the concomitant rhetorical alignment of habitation and sexual ravishment is problematised in regard to Drayton's text by the fact that Britain was not a blank space with the potential to be inscribed. In his essay 'Of Plantations', the writer, humanist philosopher and politician Francis Bacon notes: 'I like a plantation of a pure soil, that is, where people are not displanted to the end to plant in others. For else, it is rather an extirpation than a plantation'[32] This tension between plantation and extirpation on 'unpure soil', exemplified by the prolonged and troubled English colonisation of Ireland, presents a dilemma for considerations of *Poly-Olbion* and its cartographic and chorographic gendering of the British landscape. The country of England, of which Drayton was a native, could never be figured as virginal in the colonial sense. *Poly-Olbion*'s representation of a feminised Britain encompasses an apparent contradiction between the geographical literary tradition it invokes – concerned with liminal colonial spaces and outwardly empty landscapes – and its subject – a poem about the accepted imperial centre of Britain. To resolve this paradox, my analysis attempts to contextualise *Poly-Olbion* within Drayton's own biography and, in particular, his relationship to the shifting political dynamics of the early seventeenth-century court. In doing so, it reiterates the importance of geographical description as a means for political expression. Turning the feminising gaze to the autochthonous landscape, Drayton enacts an attempted possession of the land, in this case for the late monarch, Elizabeth I.

As well as considering geographical texts, I interrogate literature about mapping to scrutinise the underlying impulses of what the influential pedagogical theorist Thomas Blundeville calls 'skill in Geography'.[33] A discernible trend in such treatises is the stress on the role of cartographic knowledge in a child's educational development. This study looks at how early modern educationalists and educational theorists engaged with the map and conceptualised its function as a pedagogical aim. In doing so, I explore some of the definitional complexities surrounding the use of the term 'geography' in Renaissance humanist education, particularly in England. Through close readings of educational treatises such as Thomas Elyot's *The Boke Named the Governour* (1531), Cuningham's *The Cosmographical Glasse* (1559) and Henry Peacham's *The Compleat Gentleman* (1622), I attempt to show how some of the key ideas of the emergent 'cartographic enthusiasm' among sixteenth- and seventeenth-century intellectuals, alongside more profound notions of humanist epistemological enquiry influenced the way maps, globes and mapping instruments were used in the early modern learning environment.

While Speed and Drayton exemplify the anatomy/mapping intellectual cross-pollination in the practicalities and processes of imaging the land in the early modern period, the actual exercise of surveying, detailing and describing geographical space carried with it an elaborately structured binary of theoretical learning and practical endeavour. This combination of theory and practice is partially adumbrated by the aforementioned translation of Camden's *Britannia*: 'the best directions in Geographicall studies [occur] especially when the light of learning is adjoyned to the speechlesse delineations'. The closeness of the relationship between what Camden eloquently describes as the interaction of 'Geographicall studies' and 'speechlesse delineations' in the 'light of learning' is reiterated time and again in discussions of cartography. Thus, just as the burgeoning cartographic enthusiasm of early modern England engendered multifarious textual forms, so the varied possibilities of cartographic perspectives percolated into theories regarding knowledge acquisition and pedagogy. As this study will hope to show, writers emphasising the theory/practice paradigm at the heart of successful geographic learning anticipate the later seventeenth-century stress upon empiricist inquiry in another aspect of education and knowledge.

A leading figure in this trend was Francis Bacon. In devoting a chapter to his reflections on maps, charts and globes and their place, I explore what his ruminations on the practice of mapping and geographical description reveal about his perceptions of cartography. Through a study of Bacon's references to geography in *The Advancement of Learning* (1605) and *New Atlantis, A Worke unfinished* (1627), I endeavour to show how geographic disciplines, and mapping in particular, were used by Bacon as models for his highly influential theories on scientific practice.

Allied to these theoretical paradigms of geographical education was the practical employment of cartography and mapping in the act of discovery. Alongside new mappers of the native and known, cartographies began to appear of the recently discovered. In the period under analysis, demarcated roughly by Christopher Columbus's voyages to the New World (1490s) and the exploration of the South Pacific by William Dampier (1690s), new lands to be mapped were emerging as once empty spaces on the globe were being explored and inscribed by English and British explorers. Consequently, discovery co-existed with and often necessitated invention – new previously unknown topographies were not only uncovered, but they were also given an image fashioned and established by the discoverers. Maps provided one medium of familiarising, ordering and systematising the seemingly alien, unknowable and chaotic. 'Maps', writes Jerry Brotton in an important observation, 'were [...] regarded as rhetorically powerful tools with which to move and shape people and ideas'.[34] Cartography was commandeered as a means of producing, projecting and controlling images of the land and those who inhabited it. This was recognised by both map readers and map-pers alike: for example, John Speed's map of Ireland in *The Theatre* includes pictogram representations of the country's inhabitants – 'gentle', 'civill' and 'wilde' respectfully – to project a carefully hierarchised society. With regard to administrators and map-readers, the words of Robert Beale, secretary to the Privy Council reveal much about the understood importance of a car-tographic image of the land for its successful governance:

> A Secretarie must [...] have the booke of Ortelius' Mapps, a booke of the Mappes of England, w[i]th a particular note of the divisions of the shires into Hundreds, Lathes, Wappentaes [wapentakes], and what Noblemen, Gent[lemen] and others be residing in every one of them; what Citties, Burrows, Markett Townes, Villages; [...] and if anie other plotts or maps come to his handes, let them be kept safelie.[35]

Beale's stress on the need for the safekeeping of 'plotts' and 'Mappes' is revealing. Firstly, it shows the high status of maps in the sphere of Renaissance politics. Additionally, it underscores the perceived importance of mapped representations of space in the execution of political administra-tion. Another important example of this is the maps of William Cecil, chief advisor to Elizabeth I. Cecil's maps of Lancashire, for instance, were depic-tions not just of topography but also administrative structures, architecture and religious provision across the county. As Michael Shannon and Michael Winstanley have shown, such a cartography was a working document of government, a palimpsest of corrections, alterations and annotations as the English government sought national consolidation in the face of the Armada and Continental Catholic Europe.[36]

In the context of England and the emerging British state, the nexus of exploration, invention and mapping was imbued with colonially inflected economic intentions. The period 1500–1700 saw a massive increase in Britain's imperial ambitions.[37] Consequently, according to Bruce McLeod, 'England, on the eve of the seventeenth century was a society both in tumul-tuous flux and officially inscribing itself as its state consolidated power'. At the same time, this strengthening of state power sought to appropriate the investment of individual enterprise to further its own imperial ambitions. 'England's commercial and landed elites invested their money, learning and energy', continues McLeod, 'into the control of global trading routes and overseas expansion, subscribing to an ideology of adventure'.[38] At the centre of this endeavour was the application and use of cartography within the exercise of colonial power. Geographical science constituted a chief component of the self-perpetuating imperialist discourse in Elizabethan England. Providing a framework within which national and ethnic identities could be described, classified and hierarchised, it carried with it an unspoken substantiation of cultural supremacy. As Lesley Cormack writes:

> Geography supplied the many students and politicians who studied it with a belief in their own inherent superiority and their ability to control the world they now understood. Indeed, the study of geography helped the English develop an imperial world-view based on three underlying assumptions: a belief that the world could be measured, named, and therefore controlled; a sense of the superiority of the English over peoples and nations and thus the right of the English nation to exploit other areas of the globe; and a self-definition that gave these English students a sense of themselves and their nation.[39]

As the capitalist enterprise of early modern England and later Britain developed and colonies emerged from Ireland to Virginia to the Caribbean, mappers accompanied explorers, navigators and sailors across vast swathes of the undiscovered earth to enact their roles in the colonial endeavour. Recording the landscape, they also produced images tailored to the demands of colonial administration. This course of epistemic manufacture and the role of cartography in its function has been theorised by, among others, Michael Dillon, John Morrissey and William J. Smyth.[40] In his brilliant study *Map-Making, Landscapes and Memory* (2006), Smyth shows how, while 'Ireland is a very old land' and 'the forging of [its] physical geography is a story that involves millions of years of earth formation',[41] the impact of the English intervention, reconfiguration and re-mapping of the country particularly from the sixteenth century onwards was seismic:

> This forging involved the deformation of many older regional societies. The cumulative consequences of the continued English/British conquest, plantation and colonization was the stripping of regional lordships of many key institutions and leadership positions. Irish systems of landownership and occupation were itemized, ridiculed and eroded.[42]

Ireland became the subject of a newly created geographical image of concretised county and province boundary shapes. Here the land was parcelled into hitherto unknown regions and taxonomised under an imported English county system. In order to familiarise an alien landscape, cartographers projected a familiar geographic system onto the unknown terrain, in the process denuding existing native geographic features. Concomitantly, these new territories, demarcated and standardised by English colonial officials, manifested a whole array of political, legal and civic systems of administration.[43] The colonial impulse towards eroding and ultimately displacing the native with more recognisable and more manageable borders was part of an 'itemization' of established customs, society and landscape. Imposition of power and control went hand in hand with the imposition of a recognisable, standardised image of the land. To paraphrase Donne, early modern English mappers in Ireland such as Robert Lythe and Josias Bodley sought, at the instruction of their superiors, to capture the uncharted and unknown Ireland in a cartographic net of meridians and parallels.

The intricate and indelible relationship between mapping and colonial endeavour is examined in a chapter which analyses the role of cartography in Edmund Spenser's explicitly colonialist text *A View of the Present State of Ireland* (1633). As Andrew Hadfield's biography has shown, Spenser embarked to colonial Ireland at the age of 20, seeking the fortune and status inaccessible to him in his native land.[44] In his dialogue, we are offered a first-hand glimpse at cartography's function in the relationship between coloniser and colonised, the exercise of early modern colonial administration, and also its role in military planning as Spenser skilfully interweaves the rhetoric of cartography, colonialism, and politics in one eye-catching map reading. In laying out the map, the dialogue also flags up the purpose of mapping within the discourse of colonial power.

Another notion that emerged in Renaissance discussions of cartography, and indicating the discipline's malleability as a representational form, was the world of the map. If charts, surveys and plots could be used as tableaux for governmental function, they could also be used to fuel the imagination. Donne's figuring of cartography as a net encapsulated the subtle yet potent subtexts of possession and control at work when mappers sought to describe the land. However, there was additionally a recurring trend in discussions of cartography for conceiving of the map as a space in itself. For many Renaissance writers like Donne, who were intrigued by the map's ability to capture and corral the land, the cartographic document represented more than a reproduction of an external topography – it contained its own, internalised three-dimensional space, affording the possibility of armchair or vicarious travel.

Primarily, this belief drew its impetus from the seeming progression in mapping towards a greater accuracy and realism. 'Cartographic accuracy', writes Donald K. Smith, 'began to carry with it a concomitant sense of perceived space, a sense of an implicitly physical volume that could be imaginatively inhabited'.[45] However, it also drew strength from the belief in the power of the imagination and the mind, further revealing how cartography was alert to wider trends of intellectual ideas. The creative faculty was increasingly celebrated by Renaissance thinkers as indicative of a more substantial celebration of the individual self.

One of the most evocative contemporary literary articulations in the belief in the power of the imagination, and as a corollary, the agency of the self, appears in John Milton's *Paradise Lost* (1667). Cast down from heaven by God after his ill-fated rebellion, Satan engages in counsel with his fallen angels. In his first speech in the poem, he addresses his lieutenant Beelzebub, acknowledging their failure:

[I]f he whom mutual league,
United thoughts and counsels, equal hope
And hazard in the glorious enterprise,
Joined with me once, now misery hath joind
In equal ruin; into what pit thou seest
From what height fallen, so much the stronger proved
He with his Thunder: and till then who knew
The force of those dire Arms?
                              (Book 1, 87–94)[46]

Together in wretchedness, the archangel and his subordinate are 'joined [...] in equal ruin' in bemusement at the 'pit' of their prison. Yet this apparent desolation is soon challenged by Satan. Later in the same speech, he announces:

What though the field be lost?
All is not lost; the unconquerable will,
And study of revenge, immortal hate,
And courage never to submit or yield—
And what is else not to be overcome?
That glory never shall his wrath or might
Extort from me.
                              (Book 1, 105–111)

Satan's question 'What though the field be lost?' carries with it in its most obvious sense a military association. The 'field' to which Satan alludes, given the speech situation, is suggestive of 'battle'. Yet, as it does now, 'field' also carries with it a connotation of a geographical space or tract of land. Later in Book 1, the poem describes the construction of an edifice in Hell, evoking the physical aspect of the word: 'As when bands | Of pioneers with spade and pickax arm'd | Forerun the royal camp, to trench a field' (Book 1, 675–677). Within the speech, there is a shift in focus between two different geographies. Satan seeks to remedy the irreconcilable injury in the real space through recourse to the cerebral space. Concrete losses will be recompensed by a retreat into another sphere or 'field': 'All' Satan tellingly announces 'is not lost'. The fallen angels may retreat to the 'unconquerable will', the 'study of revenge', 'immortal hate' and 'courage'. These abstractions at once seek to establish Satan's autonomy from the tyrant God while also establishing the 'field' on which this new battle will be fought: the mind.

The mind's boundless potential and infinite creative ability is expressed in one of the most famous lines of *Paradise Lost*: 'The mind is its own place, and in itself | Can make a Heaven of hell, a hell of Heaven' (Book 1, 254–255). In this axiom, Satan underlines the perceived capability of the internal self to create its own 'place'. Framed by reference to the geographical territories of heaven and hell – territories which are all too real in Milton's epic – Satan's assertion breaks down the barrier between the imaginative and the real. The landscapes of exteriority and interiority lose their distinction as the mind becomes 'its own place'.

Evoking the dual notion of 'place' – psychological state as well as geographical area – Milton's Satan locates at the heart of his proclamation a spatialised terrain. In doing so, he draws geographical perception into concert with the power of the imagination. The potential for such a manoeuvre featured prominently in debates regarding cartography and the process of making and reading a map. Map enthusiasts of the age drew cartography into wider prevailing theories of the mind and its properties to develop the notion of vicarious travel where the map-reader might travel through the map while firmly and safely ensconced in their armchair. In a fusion of technology and the prevailing understanding of the mind's creative faculties, Renaissance theorists of the map conceived of the cartographic document as a place wherein the map-reader could, in the words of Robert Herrick, 'securely sail' ('A Country Life: To His Brother, M. Thomas Herrick', 78).[47] The recurrent presentation of this idea – from the plays of Shakespeare and Marlowe through the poetry of Donne and Herrick to the educational treatises of Cuningham and Peacham – and its accompanying implications for our understanding of how early modern culture used the maps will form a key focus of this study. Analysing how the world of the map adopted and adapted prevailing theories of mental cognition and creation, I explore how this perception of the map as a locus of a three-dimensional world was employed for educational, recreational and dramatic purposes.

Mindful of Harley's maxim that '[m]aps always show more than an unmediated sum of a set of techniques',[48] this study explores how early modern England and Britain engaged with the ideas, processes and products of cartographic practice. In doing so, it follows in a rich tradition of scholarship: the Renaissance has proved a particularly fecund area for cartographic historians who go beyond a merely artefactual consideration of the map's physical properties to ask what it reveals about wider discourses at play in the period. In the past few decades, Richard Helgerson, J.H. Andrews, Bruce Avery, Lesley Cormack and Jerry Brotton have become increasingly attentive to the socio-political and cultural dynamics underpinning geographical disciplines and their products, most especially the cartographic document. Helgerson's ground-breaking article 'The Land Speaks: Cartography, Chorography, and Subversion in Renaissance England' discusses the central role of images of the land in the formulation of national identity and communitarian consciousness.[49] By examining English cartographic representations of Ireland, Avery, William J. Smyth and Annaleigh Margey and others have drawn attention to the role of maps in the pursuit, exercise and normalisation of colonial control.[50] Elsewhere, Cormack and Jonathan M. Smith have foregrounded the role of geographic science in the formation of English imperial endeavours, and shown how the science of world-describing played a fundamental role in the cultivation of the officer class that would advance the cause of the budding empire across all four corners of the globe.[51] More recently, the work of Chris Barrett and Valerie Traub has elucidated the ways in which the cartographic gaze encompassed prevailing anxieties about nationhood and gender in a period of socio-political flux.[52]

I consciously follow in the footsteps of these scholars in attempting to get underneath the veneer of mapping's ostensibly neutral mode of representation to expose its inner workings and motivations. In the spirit of Harley's interdisciplinary scope, this book incorporates a wide variety of different textual forms and genres. It pays particular attention to mapping moments in texts which has been afforded little attention by cartographic historians including Thomas Middleton's *The Puritan* (1607), while also exploring the geographical writings of major authors such as Francis Bacon. Didactic manuals, mathematical treatises, engravings, pedagogical tracts, colonial narratives and utopian fiction are all considered with a view to understanding how early modern English culture engaged with the map and its physicality and production.

The elevation of maps as cultural objects engendered a proliferation of artistic representations of cartography and cartographic study as in poetry, plays and painting. Predominantly literary texts – plays, poems, fictional prose – are considered alongside those mentioned. Figures examined are both canonical – Shakespeare, Marlowe, Bacon – and also more marginal, for example, the émigré engraver and artist Wenceslaus Hollar. Maps feature in many different spheres of early modern culture. By drawing together just some of these instances, I will ask if a narrative of early modern cartography can be developed with its own distinct themes and methodologies of representation. And if so, how does this narrative manifest itself in cultural and socio-political discourses as conceptualisations of cartography and its potentialities percolated through the popular intellectual consciousness?

Notes

1. All quotations are taken from John Donne, *The Major Works*, ed. John Carey (Oxford: Oxford University Press, 1990).
2. For an informative account of star maps in Renaissance European culture, see Anna Friedman Herlihy, 'Renaissance Star Charts', *The History of Cartography, Volume Three, Part 1: Cartography in the European Renaissance*, ed. David Woodward (Chicago: Chicago University Press, 2007), pp. 99–122.
3. Jonathan Sawday, 'Towards the Renaissance Computer', *The Renaissance Computer: Knowledge Technology in the First Age of Print*, ed. Jonathan Sawday and Neil Rhodes (London: Routledge, 2002), p. 34.
4. Alma B. Altizer, *Self and Symbolism in the Poetry of Michelangelo, John Donne and Agrippa D'Aubigne* (The Hague: Springer, 1973), p. 91.

5. Dennis M. Lensing, 'Postmodernism at Sea: The Quest for Longitude in Thomas Pynchon's *Mason & Dixon* and Umberto Eco's *The Island of the Day Before*', *The Multiple Worlds of Pynchon's Mason & Dixon: Eighteenth-Century Contexts, Postmodern Observations*, ed. Elizabeth Jane Wall Hinds (New York: Camden House, 2005), p. 128.

6. Lorraine Daston and Peter Galison, 'The Image of Objectivity', *Representations*, 40 (1992), p. 123.

7. As J.H. Andrews playfully notes:
   The peculiar quality of Harley's achievement may best be appreciated by inventing a game for map historians: Tabulate the specialized vocabularies of cartography, philosophy, and the aggregate of all other academic subjects in three columns. Choose one term at random from each column. Whoever combines these terms into the most arresting and plausible short sentence wins the round.
   See J.B. Harley, *The New Nature of Maps: Essays in the History of Cartography*, ed. Paul Laxton with an introduction by J.H. Andrews (Baltimore and London: Johns Hopkins University Press, 2001), p. 3.

8. Harley, *The New Nature of Maps*, pp. 84–107.

9. Harley, *The New Nature of Maps*, p. 88.

10. Harley, *The New Nature of Maps*, pp. 110–168.

11. Harley, *The New Nature of Maps*, pp. 170–195.

12. Christian Jacob, 'Theoretical Aspects of the History of Cartography: Toward a Cultural History of Cartography', *Imago Mundi*, 48:1 (1996), p. 191.

13. Jacob, 'Theoretical Aspects of the History of Cartography', p. 193.

14. See, for example, Peter van der Krogt, 'Commercial Cartography in the Netherlands with Particular Reference to Atlas Production (16th–18th centuries)', *La Cartografia dels Paisos Baixos* (Barcelona: Institut Cartogràfic de Catalunya, 1994), p. 73. Jan Vermeer, one of the period's most celebrated artists, included maps in many of his paintings as well as making mappers the subject of two separate portraits. For an in-depth study of the profusion of cartography in Netherlandish art of the sixteenth and seventeenth centuries, see Svetlana Alpers, 'The Mapping Impulse in Dutch Art', *Art and Cartography: Six Historical Essays*, ed. David Woodward (Chicago: University of Chicago Press, 1987), pp. 51–96.

15. Richard Helgerson, 'Introduction', *Early Modern Literary Studies* 4.2:3 (1998), para. 8.

16. John Dee, *The Elements of geometrie of the most auncient philosopher Euclide of Megara* (London: John Day, 1570), sig. A4r.

17. Thomas Blundeville, *A Briefe Description of the Universal Mappes and Cardes, and of Their Uses* (London: Roger Ward, 1589), sig. A3r.

18. Blundeville's remark has been challenged by Gavin Hollis, who points out that while many writers envisioned a wide level of map and atlas ownership and the stocking of libraries with cartographic objects, the cost of such items was often prohibitive. Writing of Ortelius's *Theatrum*, Hollis observes, '[N]ot many Elizabethans could afford libraries, nor could many afford the atlas in the first place'. See Gavin Hollis, '"Give Me the Map There": *King Lear* and Cartographic Literacy in Early Modern England', *The Portolan*, 68 (2007), p. 10.

19. Donald K. Smith, *The Cartographic Imagination in Early Modern England: Re-writing the World in Marlowe, Spenser, Raleigh and Marvell* (Surrey: Ashgate, 2008), p. 6.

20. John Norden, *Nordens preparatiue to his Speculum Britanniae* (London: Eliot's Court Press, 1596), p. 1.

21. William Cuningham, *The Cosmographical Glasse conteinyng the pleasant principles of cosmographie, geographie, hydrographie, or nauigation* (London: John Daye, 1559), p. 6.
22. William Lambarde, *A perambulation of Kent* (London: Ralph Newberie, 1576 [1570]), sig. ¶1r.
23. John M. Adrian, *Local Negotiations of English Nationhood, 1570–1680* (Basingstoke: Palgrave Macmillan, 2011), pp. 12–13.
24. William Camden, *Britain, or A chorographicall description of the most flourishing kingdomes, England, Scotland, and Ireland, and the ilands adjoyning, out of the depth of antiquitie beautified vvith mappes of the severall shires of England*, trans. Philemon Holland (London: George Bishop and John Norton, 1610), sig. ¶5r.
25. John Speed, *The Theatre of the Empire of Great Britaine* (London: George Humble, 1611), sigs. ¶3v-4r.
26. Isaiah 28: 16.
27. J.H. Andrews, 'Statements and Silences in John Speed's Map of Ulster', *The Journal of the Royal Society of Antiquaries*, 138 (2008), p. 71.
28. J.H. Andrews, 'Colonial Cartography in a European Setting: The Case of Tudor Ireland', *The History of Cartography. Volume 3: Cartography in the European Renaissance* (Chicago: Chicago University Press, 2007), p. 1682.
29. Valerie Traub, 'The Nature of Norms in Early Modern England: Anatomy, Cartography, *King Lear*', *South Central Review*, 26:1 & 2 (2009), p. 43.
30. Speed, *Theatre*, sig. ¶3r.
31. Annette Kolodny, *The Lay of the Land: Metaphor as Experience and History in American Life and Letters* (North Carolina: University of North Carolina Press, 1975), p. 4.
32. Francis Bacon, *The Essays*, ed. with an introduction by John Pitcher (London: Penguin Books, 1985), p. 162.
33. Thomas Blundeville, *M. Blundevile His Exercises* (London: 1594), sig. A3r.
34. Jerry Brotton, 'Tragedy and Geography', *A Companion to Shakespeare's Works. Volume 1: The Tragedies*, ed. Richard Dutton and Jean E. Howard (Oxford: Blackwell Publishing Ltd., 2003), p. 221.
35. Robert Beale quoted in Swen Voekel, '"Upon the Suddaine View": State, Civil Society and Surveillance in Early Modern England', *EMLS*, 4.2:3 (1998), para. 1.
36. William Shannon and Michael Winstanley. 'Lord Burghley's Map of Lancashire Revisited, c. 1576–1590', *Imago Mundi*, 59:1 (2007), pp. 24–42.
37. A.L. Rowse, *The Expansion of Elizabethan England* (Wisconsin: University of Wisconsin Press, 2003); and Nicholas Canny, 'The Origins of Empire: An Introduction', *The Oxford History of the British Empire, Volume 1: The Origins of Empire*, ed. Nicholas Canny (Oxford, Oxford University Press, 2001), pp. 1–33.
38. Bruce McLeod, *The Geography of Empire in English Literature 1580–1745* (Cambridge: Cambridge University Press, 1999), p. 37. See also Rory Rapple, *Martial Power and Elizabethan Political Culture: Military Men in England and Ireland, 1558–1594* (Cambridge: Cambridge University Press, 2009), pp. 51–85 for an account of the socio-political origins of this class of entrepreneurial colonialists.
39. Lesley B. Cormack, 'Britannia Rules The Waves?: Images of Empire in Elizabethan England', *EMLS*, 4.2:3 (1998), para. 1.
40. Michael Dillon, 'Governing Through Contingency: The Security of Biopolitical Governance', *Political Geography*, 26:1 (2007), pp. 41–47; John Morrissey, 'Foucault and the Colonial Subject: Emergent Forms of Colonial

Governmentality in Early Modern Ireland', *At the Anvil: Essays in Honour of William J. Smyth*, eds. Patrick Duffy and William Nolan (Dublin: Geography Publications, 2012), pp. 135–150 and William J. Smyth, *Map-making, Landscapes and Memory: A Geography of Colonial and Early Modern Ireland c. 1530–1750* (Indiana: University of Notre Dame Press, 2006).

41. Smyth, *Map-making, Landscapes and Memory*, p. 1.

42. Smyth, *Map-making, Landscapes and Memory*, p. 4.

43. This process, and the centrality of mapping to its execution, is summed up by Smyth: 'The shiring of so many new counties and the creation of further layers of administration at the barony and manorial court levels progressively undermined the older territorial and legal systems of rule. However, there is little doubt but that the introduction of full-blooded plantations – involving elaborate planning, detailed mapping, the introduction of new landed elites and settlers, the creation of many new plantations and villages and the intensification of a market economy – was central to this transformation' (Smyth, *Map-making, Landscapes and Memory*, p. 101).

44. For a more in-depth biography of Spenser and his time in Ireland, see Andrew Hadfield, 'Spenser, Edmund (1552?–1599)', *ODNB* (Oxford: Oxford University Press, 2004); online edn. January 2008, http://www.oxforddnb.com/view/article/26145 (accessed 2nd April 2014); and also Hadfield, *Edmund Spenser: A Life* (Oxford: Oxford University Press, 2012).

45. Smith, *The Cartographic Imagination in Early Modern England*, p. 8.

46. All quotations are taken from John Milton, *Paradise Lost*, ed. David Scott Kastan (Indianapolis, Indiana: Hackett Publishing Ltd., 2005).

47. All quotations are taken from Robert Herrick, *The Complete Poetry of Robert Herrick*, ed. Tom Cain and Ruth Connolly (Oxford: Oxford University Press, 2013).

48. Harley, *The New Nature of Maps*, p. 36.

49. Richard Helgerson, 'The Land Speaks: Cartography, Chorography, and Subversion in Renaissance England', *Representations*, 16 (1986), pp. 50–85.

50. See, for example, Bruce Avery, 'Mapping the Irish Other: Spenser's *A View of the Present State of Ireland*', *ELH*, 57:2 (1990), pp. 263–279; David J. Baker, 'Off the Map: Charting Uncertainty in Renaissance Ireland', *Representing Ireland: Literature and the Origins of Conflict 1534–1660*, eds. Brendan Bradshaw, Andrew Hadfield and Willy Maley (Cambridge: Cambridge University Press, 1993), pp. 76–92; William J. Smyth, *Map-making, Landscapes and Memory* (Notre Dame, Illinois: University of Notre Dame Press, 2006) and Annaleigh Margey, 'Representing Plantation Landscapes: The Case of Ulster, c. 1560–1640', *Plantation Ireland: Settlement and Material Culture, c. 1550–1700*, eds. C. Rynne and J. Lytellton (Dublin: Four Courts Press, 2009), pp. 140–164 and 'Visualising the Plantation: Mapping the Changing Face of Ulster', *History Ireland*, 17:6 (2009), pp. 42–45.

51. Lesley B. Cormack, *Charting an Empire: Geography at the English Universities, 1580–1620* (Chicago: Chicago University Press, 1997) and Jonathan M. Smith, 'State Formation, Geography, and a Gentleman's Education', *Geographical Review*, 86:1 (1996), pp. 91–100.

52. Chris Barrett, *Early Modern English Literature and the Poetics of Cartographic Anxiety* (Oxford: Oxford University Press, 2019). Valerie Traub, 'Anatomy, Cartography, and the New World Body' in *Geographies of Embodiment in Early Modern England*, eds. Mary Floyd-Wilson and Garrett. Sullivan, Jr. (Oxford: Oxford University Press, 2020), pp. 64–112.

1 'they say the world's in one of them'
The World of the Map

I would like to start by considering how early modern English intellectual culture approached the map as a physical, three-dimensional object. An instructive insight is found in Thomas Elyot's *The Boke Named the Governour* (1531), which, like its continental counterparts Desiderius Erasmus's *The Education of a Christian Prince* (1516), Baldassare Castiglione's *The Book of the Courtier* (1528) and Niccolò Machiavelli's *The Prince* (1532), fashions itself as a guidebook on statecraft, diplomacy and humanistic learning. In a section discussing education, Elyot reflects upon the pleasures of map reading:

> For what pleasure is it in one hour, to behold those realms, cities, seas, rivers, and mountains, that not easily in a whole man's life can be journeyed and pursued: what incredible delight is taken in beholding the diversity of people, beasts, fowls, fishes, trees, fruits, and herbs: to know the sundry manners and conditions of people, and the variety of their natures, and that in a warm study or parlor, without peril of the sea, or danger of long and painful journeys; I cannot tell what more pleasure should happen to a gentle wit, than to behold in his own house every thing that within all the world is contained.[1]

Elyot's panegyric to the 'pleasure' of map reading has been identified as illustrative of the early modern period's fascination with maps and the gratification of cartographic study.[2] The eulogising of the experiential benefits of map reading marks out several facets of the contemporary understanding of the cartographic document. Primarily, it imparts in a fulsome fashion the map's ability to gratify intellectually: 'I cannot tell', Elyot states confidently of map study, 'what more pleasure should happen to a gentle wit'. Secondly, it highlights the variety of the representational scope of the map. Alongside geography, it portrays topography, demography, mammalogy, botany and even ichthyology. The emphasis on diversity is clear: pluralised phrases such as 'sundry manners', 'conditions of people' and 'variety of natures' foreground the multifaceted character of the map.

DOI: 10.4324/9781003200376-2

Perhaps the most noticeable aspect of Elyot's description, however, is its articulation of the map's ability to contain everything in a single, easily perceivable view. When describing the cartographic document and its contents, Elyot repeatedly employs the definite article – '*the* diversity of people, beasts, fowls, fishes, trees, fruits, and herbs', '*the* sundry manners and conditions of people', '*the* variety of their natures' – suggesting an authoritative comprehensiveness to the range of information shown in the map. Alongside this suffusion of stuff is a focus on the map's contraction of space. The vast distances of 'real' geography are condensed into a single document which may be viewed in 'one hour'. The presence of a specific number here is significant: while all the other diversities named by Elyot are without numerical limit ('people, beasts, fowls, fishes, trees, fruits, and herbs'), the length of study is defined clearly, suggesting the map contains within it the ability to encompass the limitless multiplicity of the earth in a circumscribed prospect. Elyot's rhetoric adroitly counterbalances the multiplicities of the earth with the singular perspective of the map. Importantly, this contraction by the map-maker is not characterised by dilution but rather concentration. The map offers a panoramic view of the land, its natural environment and people. It provides insight into the social, cultural and historical lives of those inhabiting 'those realms, cities, seas, rivers, and mountains' by condensing all into one view everything that 'in all the world is contained'.

In reducing everything to one vista, studied in a distinguishable time frame, the map also offers to the reader a form of travel. This idea is expressed through celebrations of the cartographic enthusiast's ability to 'travel by map', or in Elyot's words, 'behold [...] the diversity of people, beasts, fowls, fishes, trees, fruits, and herbs' and 'to know the sundry manners and conditions of people, and the variety of their natures, and that in a warm study or parlor, without peril of the sea, or danger of long and painful journeys'. The use of the suggestive verb 'beholde' is emblematic of Elyot's multi-layered understanding of the map, conveying both a sense of visual perception and also personal engagement. It suggests that the early modern experience of map reading went beyond a superficial glance at a representation of geographical space, extending to a much more profound interaction with the output of the cartographer's hand.

The map and mind travel

By verbalising such an idea, Elyot tapped into broader enthusiasms. As David McInnis has shown, 'mind-travelling' was popular among early modern readers, 'whether [...] because travel anywhere was prohibitive for the mass of individuals, or because of the hardships of the journey'.[3] Cartographic science proved a fertile ground for such enthusiasms. The sentiments of Elyot's observations echo through much cartographic writing: time and again intellectuals celebrated the world of the map and cartography's apparent ability to transcend the physical boundaries of the enclosed study. Many late sixteenth- and early seventeenth-century texts that discuss cartography, or reference the study and production of maps, reciprocate Elyot's enthusiastic description of the 'pleasure' of map reading. In a similar feting of cartographic travel in his famous *Anatomy of Melancholy* (1621), Robert Burton writes:

> To some kind of men it is extraordinary delight to study, to looke upon a Geographicall mappe, and to behold, as it were, all the remote Provinces, Townes, Cities of the world, and never to goe forth of the limits of his study, to measure by a Scale and Compasse, their extent, distance, examine their site, &c.[4]

Described by Bernhard Klein as an exercise in 'exuberant praise of a mode of pictorial representation which allows the hazards of travel to be replaced by the comforts of solitary contemplation',[5] Burton's comment recognises the belief among cartographic enthusiasts ('some kind of men') regarding the map's capacity to draw together 'all the remote Provinces, Townes, Cities of the world' until they are contained safely in the 'limits' of the private study.[6] Recalling in linguistic detail Elyot's description of the 'pleasure' of maps written nearly a century earlier, Burton characterises the 'study' of the cartographic document using the same evocative term, 'behold'. This 'extraordinary delight' is marked by an exploratory experience that perceives the map as exceeding its two dimensions.

Published between Elyot's *The Boke Named the Governour* and Burton's *Anatomy of Melancholy*, and containing similar sentiments, is *The Cosmographical Glasse* (1559). Written by the physician, astrologer and cartographer William Cuningham, this technical treatise has been identified as an important text in the increasing sophistication in cartographic practices in England from the middle of the sixteenth century.[7] Written in the form of a dialogue between student and teacher, Cuningham's text is largely scientific and 'entirely concerned with technical precision in measuring land, reading the sky and producing accurate maps'.[8] The book also contains a lyricised acclamation of the pleasure of map reading which aligns the author with similar cartographic enthusiasts and further attests to this trend in the map-minded literature of the period. Cuningham, like Elyot and Burton, proclaims the map's ability to reduce the world to one single prospect. Set amidst an elaborate frontispiece depicting the ancient astronomers Ptolemy and Strabo atop heavenly clouds alongside personifications of human knowledge such as Geometria, Arithmetica and Musica, an introductory sestet lauds the comprehensive scope of the 'cosmographical glasse' or map and its unique quasi-replication of the 'real' world:

> In this Glasse if you will beholde
> The Sterry Skie, and Yearth so wide,

The Seas also, with windes so colde,
Yea and thy selfe all these to guide:

What this Type meane first learne a right,
So shall the gayne thy trauaill quight.[9]

In this short poem, the visual elements contained in Cuningham's 'Glasse' – the 'Sterry Skie' and the extent of earth or 'Yearth so wide' – are allied to another characteristic, namely the 'cold' of the 'windes'. Such a description of the cartographic study is, at first reading, challenging. Assuming a degree of map-reading ability such as familiarity with scalar measurements, the discernment of physical proportions is a fairly straightforward proposition. However, physically sensing cold and wind simply through looking at a pictorial representation of a tract of land appears outlandish. In one sense, Cuningham's description is fantastical, ascribing impossible attributes to the map. However, the introductory poem to *The Cosmographical Glasse* demonstrates the allure of the world on the map. It expands upon the possibilities of 'viewing', 'examining' and 'beholding' the world expressed by Elyot, depicting map-reading as a multi-sensual experience. The cartographic document engenders an amplificatory discourse with writers celebrating its transmission of a more comprehensive sensual experience, one which stimulates not only the eyes but a number of other senses too.

This conceptualisation of cartographic study as a form of travel is elucidated later in *The Cosmographical Glasse*:

Nowe I perceive by the makinge and describyng of this onely Mappe [...] that observing this order of you prescribed, I may in like sorte at my pleasure, drawe a Carde for Spaine, Fraunce, Germany, Italye, Graece, or any perticuler region yea, in a warme & pleasaunt house, without any peril of the raging Seas: danger of enemies: losse of time: spending of substaunce: werines of body, or anguishe of minde. Oh how precious a jewell is this, it may rightly be called a Cosmographicall Glasse, in which we may beholde the diuersitie of countries: natures of people, & innumerable formes of Beastes, Foules, Fishes, Trees, Frutes, Stremes, & Meatalles.[10]

The stylistic and thematic parallels in the respective accounts of the map study discussed are recognisable. There is a reiteration of the map's visual condensation of the world. For Cuningham, such spatial epitomisation prompts the jubilant description of the map as a 'precious [...] jewell' and 'Cosmographicall Glasse'. Cuningham, like Elyot and Burton, employs the rhetorical device of enumeratio or compact, forceful listing to evoke the cornucopia of environments contained in the map: where Elyot refers to 'the diversity of people, beasts, fowls, fishes, trees, fruits, and herbs' and Burton refers to 'all the remote Provinces, Townes, Cities of the world', Cuningham points to the map's 'diuersitie of countries' and the 'natures of people, &

innumerable formes of Beastes, Foules, Fishes, Trees, Frutes, Stremes, & Meatalles'. There is also reference to the idea of cartographic travel and the leisurely nature of the practice of cartographic study – as in Burton, the scholar takes 'extraordinary delight' to peruse the map, so Cuningham's chart ('carde') is produced and examined 'at my pleasure'. Characterising map reading as a luxurious, perhaps even extravagant activity, such descriptions reveal the degree of wonderment among scholars when viewing such objects.

The map and the study

Another notable feature in discussions of early modern cartographic study is an assertion of the circumscribed nature of the room wherein such study takes place. Cuningham, for example, refers to the comfort and pleasant nature of his surroundings as he examines the map. For his 'warme & pleasaunt house', we might easily substitute without significant alteration of meaning Elyot's 'warm study or parlor' or Burton's 'study'. Such enclosure provides a contrast to the expansiveness of the map, underlining the imaginative capaciousness of the cartographic document. Furthermore, this recurrent motif simultaneously accentuates the safety of travelling by map, as compared with an actual physical excursion. The distinction between the variegated world and the safety and warmth of the study, parlour or house reinforces the perception of the map as a means of secure travel. In Cuningham's case, in particular, the articulation of the temperateness of where the map is being studied contrasts with the 'windes so cold' mentioned in the introductory verse to underline the sensual experience of map travel. Once more juxtaposition – the circumscribed map sheet as against the expansive imaginative cartography, the enclosure of study versus the liberation of the creative mind inspired by the map – is employed to indicate the enthusiasms that lie at the heart of the period's cartographic consciousness.

The writings of the surveyor John Norden continue this rhetorical tradition to foreground the conceptualisation of the map as a unique spatial construction in dialogue with the imagination. *The Surveyor's Dialogue* first published in 1607, is one of the most important texts in the evolution of early modern mapping methods and systems.[11] Concerned largely with the practice of map-making and its role in the socio-economic relations of early seventeenth-century agrarian England, the *Dialogue* takes, as the title suggests, the form of a discussion between a surveyor and a concerned farmer whose land is being surveyed. When the farmer asks why his landowner has employed the surveyor, the surveyor sets out the purpose of his task. He aims to produce:

> a plot rightly drawne by true information, describ[ing] so the lively image of a Manor, and every branch and member of the same, as the Lord sitting in his chair, may see what he hath, where, and how it lieth, and in whose use and occupation every particular is, upon the suddaine view.[12]

The surveyor's task is to construct a space in which the lord or landowner can journey through his land from the confines of his 'chayre' and 'see what he hath, where and how it lyeth'. The map for Norden's surveyor, as for others, facilitates a form of travel through the estate.

Additionally, in Norden's description, the regard of the map extends beyond the merely topographical to encompass a wider, more varied survey of the tenants and their 'use and occupation'. Among the celebrations of cartography's efficacy examined so far, *The Surveyor's Dialogue* represents the clearest suggestion of the map as a means of encompassing what we might now call 'human geography'. According to John Agnew and David N. Livingstone, this area of geographical science is concerned with 'the relationship between the natural and the human worlds, the spatial distributions of human phenomena and how they come about, the social and economic differences between different parts of the world'.[13] Thus, where Cuningham appends to the visual facet of the map a sensual awareness of the natural environment, of wind and heat, Norden imbues the map with yet another imperative – the overt projection of social environment and socio-economic relations. The notion that map readers could grasp a more profound understanding of the lives of the inhabitants of the geography represented in the map was, as we have seen, a persistent theme – witness, for example, Elyot's suggestion that the cartographic enthusiast has an insight into the 'the sundry manners and conditions of people'.

However, placed in the context of the power dynamic of Norden's surveyor/tenant discourse, this aspect of mapping is imbued with special significance. The survey, with its all-encompassing outlook, becomes a means of surveillance. The otherwise concealed lives and dealings of the tenanted population and their usage of the land – 'the Lord [...] may see what he hath, where and how it lieth, in whose use and occupation' – are made immediately accessible to the observer through the medium of the map. The survey and its function are not just concerned with the geographical features of the land; it is also capable of relating and revealing the intricate sociological and economic inter-relations among the people who inhabit the land.

The map described by Norden's surveyor anticipates a more recent understanding of the map as a projection and validation of existing social structures and attests to the nuanced complexities of the map's character in the early modern imagination. Maps, J.B. Harley writes, are not merely 'inert records of morphological landscapes or passive reflections of a world of objects'; but instead 'are regarded as refracted images contributing to dialogue in a socially constructed world'.[14] In essence, the perspective of the 'plot' in Norden's dialogue encompasses both geographical and social realities, inscribing, containing and articulating a multivalent picture of the land and its inhabitants.[15]

In this process, Norden's map adds another dimension to the early modern map, namely the discourse of power. Through its depiction of an object which the landowner might use to 'travel' through his land, it reiterates the perception of the map as its own world. Yet, by incorporating the 'use and occupation' of the tenants, the cartographic document acts as both a reproduction of a topographical environment and also the reinforcement of an established socio-economic schema. The surveyor repeats the alleged impartiality of the exactitude and accuracy of cartography – 'a plot rightly drawne by true information'. Yet while the surveyor is able to present the world as it seemingly appears, he also calls on the notion of the map's understood ability to project a world to buttress the existing economic interrelations between tenants and landlords. The map is, at first, a seeming presentation of reality and, at the same time, a validation of current norms and structures. Significantly, through the striking phrase 'rightly drawne', the surveyor of Norden assumes an authoritative stance on both accuracy and purpose.

Imaginative cartographies: Herrick and Donne

The map's ability to evoke its own world, maintained in Norden and others, is reciprocated by contemporaneous poetry. A lyrical iteration of this theme is found in 'A Country Life: To His Brother, Mr. Thomas Herrick', written in 1610 by Robert Herrick.[16] Published later in the collection *Hesperides* (1648), the poem promotes the ideal of simple country existence against the corruptions of city and courtly life. An address, as the title suggests, to the poet's brother ('Thrice, and above, blest, my soul's half, art thou', 1),[17] the lyric extols the virtues of country living and 'the country's sweet simplicity' (4). Framing the natural world as a locus of virtue (15–20) and peaceful harmony (41–50) and a teacher of 'ways less to live, than to live well' (10), Herrick repeats the idealisation of the rural environment that recurs frequently in pastoral literature. Moreover, it articulates a poeticised cartographic world: much like those discussed previously, Herrick's belief in the world of the map is intimated through the delineation of map study as a form of exploration and travel. Among the many pleasures of rustic life, the poem includes the perusal of maps: while asserting his brother is free from 'the desp'rate cares | Th'industrious merchant has' (67–68), Herrick writes:

> [...] thou at home, blest with securest ease,
> Sitt'st, and believ'st that there be seas,
> And watery dangers; while thy whiter hap
> But sees these things within thy map;
> And viewing them with a more safe survey,
> Mak'st easy fear unto thee say,
> 'A heart thrice walled with oak and brass, that man
> Had, first durst plough the ocean'.
> But thou at home, without or tide or gale,
> Canst in thy map securely sail;
> Seeing those painted countries, and so guess

By those fine shades, their substances;
And from thy compass taking small advice,
Buy'st travel at the lowest price.

(66–85)

Like the cloistered scholar poring over a map portrayed in Elyot, Cuningham and Burton, or Norden's enthroned lord scrutinising a survey of his lands, Thomas Herrick is able to experience the world as though he were physically travelling through it. The topography of the map within his circuit, he can explore in safety (76), security (81) and with ease (72) and frugality (85) the 'painted countries' (82). Invoking one of the major themes of the early modern cartographic enthusiast, Herrick's poem portrays map study as a type of worldly navigation and exploration. The line between physical travel and imagined travel is blurred through several techniques, especially the repeated use of verbs. The many references to action (seeing, viewing, making, ploughing, taking, sailing, guessing, buying) in the poem vividly distinguish the vibrancy of mind travel via the cartographic document from the 'real' sedentary state of the addressed. Though 'thrice walled with oak and brass' and engaged in the solitary act of map study, Thomas Herrick is portrayed in a state of perpetual motion, carrying out a wide range of activities.

In addition, the rhetoric of juxtaposition is repeated as the poet draws a direct parallel between the aesthetic characteristics of the map and the physical characteristics of the 'real' world: going beyond the obvious spatial correlation between cartography and topography, his brother is able, through examining the 'fine shades' of the map, to 'guess' or surmise the 'substances' of those lands contained therein. The word 'substances' in this context evokes a sense of physicality that exceeds the immediately visually perceivable and gestures towards a more profound realisation of the constitution of the places depicted. Herrick shares Cuningham's emphasis on the multi-sensual experience of map study as well as Norden's articulation of the cartographic document's ability to contain not just topographical features but also human geography in delineating a multi-dimensional model of the map. The brotherly assurance that 'thou at home [...] canst in thy map securely sail' repeats the idea of travelling by cartographic study and as a result, locates Herrick within a broader tradition of cartographic enthusiasm, which propagated the idea of the world of the map.

This paradigm finds its most lyrical enunciation in the work of John Donne. Donne's writing, as Lisa Gorton notes, 'shows he was fascinated by new discoveries [and] [h]e took up the modern idiom of maps and discovery with delight' and we have seen in the introduction how Donne skilfully uses cartographic imagery.[18] Some of the poet's most famous conceits employ the language of cartography and geography, from the 'stiff twin compasses' (25–28) of 'A Valediction: Forbidding Mourning' to the femininised 'new-foundland' (27) of 'Elegy XIX'.[19] The cartographic world is also delineated in 'The Good Morrow'. It is, like Herrick's lyric, fashioned as an explicitly private poem and cartography is used as a point of contrast to stress the delights of one of the poet's most frequent themes – the erotic encounter.[20] Donne's lyric is amatory as opposed to fraternal in tone, and the opening stanza establishes the intimate speech situation and singular addressee:

> I wonder, by my troth, what thou and I
> Did, till we loved? Were we not weaned till then?
> But sucked on country pleasures, childishly?
> Or snorted we in the Seven Sleepers' den?
> 'Twas so; but this, all pleasures fancies be.
> If ever any beauty I did see,
> Which I desired, and got, 'twas but a dream of thee.
>
> (1–7)

The speaker, insistent upon the power of love to 'make one little room an everywhere', tells their paramour:

> Let sea-discoverers to new worlds have gone,
> Let Maps to others, worlds on worlds have shown,
> Let us possess one world; each hath one, and is one.
>
> (12–14)

Pleading to the addressed to affect an ignorance of the outside world, these anaphoric implorations of the speaker are rendered increasingly complex the more they are scrutinised. A case in point is the repetition of the word 'world' – occurring four times in three lines (three as plural and once as singular), it does not appear to be the subject of a pun (a recurrent stylistic feature of Donne's poetry) and seems, at first sight, consistent in its denotation of a geographic space. In the first instance, the 'new worlds' of the 'sea-discoverers' are comprehendible: beyond the intimacy of the speaker and the audience, these spaces are to both novel and unfamiliar, perhaps signifying the New World whose exploration occupied many Western voyagers in the early modern age. They are the concern, suggests Donne, not of the lovers but of 'sea-discoverers', privateers and explorers.

However, this singular of meaning quickly dissipates as the poem's 'worlds' multiply. 'Let Maps to others, worlds on worlds have shown' urges the speaker, requesting the addressee to fix their gaze upon the private, intimate space. Repeating the plea contained in the previous lines to ignore the outside world, Donne also introduces another space into the poem, namely the map. The physicality of the cartographic document is foregrounded – it is its own world, a physical space encompassing another physical space, rather than a gridded abstraction or representation of the 'real'.

The quadripartite configuration of 'The Good Morrow', wherein worlds are piled on worlds, expanses upon expanses reminds us more generally of the intricate style of a poet who revelled in paradoxes, wordplay and the richness of figurative language. It also brings to the forefront Donne's intricate and multivalent engagement with space, both mapped and imaginative. In relation to cartography and the cartographic document, Donne's adroit use of paronomasia – or repetition of a similar word in a different sense – reiterates the prevailing conviction that somehow a map can contain a world. The world is both shown on the map, and the map is its own world.

Superficiall figures: The magic of the map

Thus far, we have examined broadly celebratory discussions of the map. Cartographic enthusiasts such as Burton, Norden and Donne expound and expand upon the established idea of the 'cosmographical glass' espoused by Elyot and Cuningham. In Herrick and Donne, the popularity of this idea is evident in lyrical allusion to 'vicarious travel'. Yet there was also an antithesis to this effusive discourse of cartographic enthusiasm, one which pointed out the representational limits of the map, even connecting cartography to more nefarious practices such as black magic. In his prose work *A Defence of Ryme* (1603), the writer Samuel Daniel warns his reader:

> We must looke upon the immense course of times past as men overlooke spacious and wide countries from off high Mountaines, and are never the neere to judge of the true Nature of the soyle or the particular syte and face of those territories they see. Nor must we thinke, viewing the superficiall figure of a region in a Mappe that wee know strait the fashion and place as it is. Or reading an Historie (which is but a Mappe of Men, and dooth no otherwise acquaint us with the true Substance of Circumstances then a superficiall Card doth the Seaman with a Coast never seene, which always proves other to the eye than the imagination forcast it), that presently wee know all the world, and can distinctly judge of times, of men, and maners, just as they were.[21]

As Bernhard Klein surmises, for Daniel, maps 'are, in fact, instruments of worrying delusion which may effectively prevent visual access to vital information by keeping the viewer at a distance from the "reality" of the region depicted'.[22] Daniel warns against any supposition of insight on the part of the map reader. 'Nor must we thinke viewing the superficiall figure of a region in a Mappe that wee know strait the fashion and place as it is'. The critique of cartography in *A Defence of Ryme*, with its nod towards the inherent abstraction of cartographic representation, presents a counterpoint to the delighted map enthusiasts explored previously. Maps are by necessity not the real, and the reader should be careful not to believe them adequate substitutes, regardless of their seeming accuracy. To place faith in such simulacra, Daniel argues, would be irrational.

This scepticism towards cartography was repeated by other writers and is featured in a play entitled *The Puritan, or the Widow of Watling Street* (1607). Intriguingly attributed to one 'W. S.' though now widely accepted as the work of Shakespeare's occasional collaborator Thomas Middleton, the play presents us with two sergeants-at-arms who attempt to apprehend the play's anti-hero, penniless scholar George Pieboard for an unpaid debt.[23] As part of a diversion to facilitate a getaway, Pieboard hands his captors 'Maps, and prittie toyes', provoking the anxious officer Puttock to observe:

> These Mappes are prittie painted things, but I could nere fancie 'em yet, mee thinkes they're too busy, and full of Circles and Conjurations, they say the world's in one of them, but I could nere find the Counter in the Poultrie.
>
> (3.3. 102–106)[24]

Puttock acknowledges the popular belief in the world of the map ('they say the world's in one of them'). However, his guardedness towards maps – 'mee thinkes they're too busy', a pejorative adjective synonymous in the early 1600s with 'prying' or 'meddlesome'[25] – also focuses on their suspicious 'Circles and Conjurations'.

The use of 'conjuration' to describe the contents of maps is a particularly potent objection. According to the *Oxford English Dictionary*, the term had a number of meanings in the early modern period, including 'banding together by oath'; and also 'the compelling of spirits or demons, by such means, to appear and do one's bidding' or 'A magical form of words used in conjuring; a magic spell, incantation, charm'. This latter definition was widespread and frequently occurred in religious discourse. Protestant polemicists often employed the term 'conjuration' and its verb 'conjuring' to attack Catholic religious customs, for example, the belief in transubstantiation during the sacrament of the Eucharist. In a collection of his work published posthumously in 1573, the Protestant theologian William Tyndale differentiates between the Protestant 'Christes signes or Sacramentes' and the Catholic 'Antichristes signes or ceremonyes' or 'disguisinges and Apesplay, and of all maner conjurations'.[26] Corroborating Tyndale's association of Catholic liturgy with conjuring (and attesting to its persistence in the seventeenth century) is Thomas Hobbes. Writing in his treatise on civic society *Leviathan* (1651), Hobbes invokes the term in the penultimate section entitled 'Of the Kingdom of Darkness' to inveigh against 'the second generall abuse of Scripture, [that] is the turning of Consecration into Conjuration, or Enchantment'. Hobbes writes:

> [W]hen by such words [the Catholic liturgy], the nature or quality of the thing it self, is pretended to be changed, it is not consecration, but either an extraordinary worke of God, or a vain and impious conjuration. But seeing (for the frequency of pretending the change of nature

in their consecrations,) it cannot be esteemed a work extraordinary, it is no other than a conjuration or incantation, whereby they would have men to believe an alteration of nature that is not, contrary to the testimony of man's sight, and of all the rest of his senses.[27]

Hobbes's usage exemplifies the negative connotations surrounding 'conjuration' in the sixteenth and seventeenth centuries. Hobbes, like Tyndale, associates 'conjuration' with dubious and even ungodly practice: 'vaine' and 'impious', 'Conjuration' is a trick which seeks to fool the senses and 'would have men to beleeve an alteration of Nature that is not, contrary to the testimony of mans Sight'.

Thus, Puttock's objection presents a negative perception of cartography, aligning it with suspicious religious practices critiqued by Protestant reformers. The celebrated ability of maps to seemingly contain a world is subverted in Puttock's rhetoric as its representational potential is yoked to the dubious and even ungodly practice of conjuring. Such an association would have contemporary resonance: in the immediate context of *The Puritan*'s first performance (1606–1607), conjuration and its links with black magic was a topic that provoked fascination, intrigue and dread.[28] The reigning monarch, James I, was renowned for his interest in the occult, composing a tract on the subject entitled *Daemonologie* (1597). In the theatre, the first appearance of Middleton's work roughly coincided with the appearance of William Shakespeare's *The Tragedy of Macbeth,* which stages extensive scenes of witchcraft. Furthermore, upon hearing the line 'but I could nere fancie 'em yet, mee thinkes they're too busy, and full of Circles and Coniurations', the early modern theatregoer may have been reminded of the most famous occultist of the English Renaissance stage, *Doctor Faustus.* In the first scene of Christopher Marlowe's 'tragical history', the main character Faustus bids farewell to the dictums and propriety of 'Divinity'. Picking up, according to stage direction, his 'book of magic', the protagonist greets them with the words:

> These metaphysics of magicians
> And necromantic books are heavenly,
> Lines, circles, signs, letters, and characters-
> Ay, these are those that Faustus most desires.
> (1.1. 51–53)[29]

Similarities between Puttock's description of the map and Faustus's acclamation of his own occult tome are conspicuous. Both reference the accumulation of things on the page: Puttock complains that the map is 'too full' while Faustus is provoked by the abundance of figures before him to an enumerative extolling of the contents of the book: 'Lines, circles, signs, letters and characters'. Additionally, both describe the appearance of 'circles', suggesting a similarity between Puttock's map and the books used by

the occultist for his magic. The striking similarities with the use of enumeration by the likes of Elyot and Cuningham in their paeans to cartography
demonstrate how the map's representational potential could be seen as both
edifying and perilous.

Middleton's character was not alone in articulating a connection between
black magic and the map. The suspicion of the serjeant-at-arms finds a like
mind in the writings of the divine William Cowper. In his exegetical tract
Three Heavenly Treatises, Concerning Christ (1612) Cowper refers to cartography in an analysis of Jesus's ascent of the mountain where he is shown 'all
the kingdoms of the world' and tempted by Satan.[30]

> How it is that Sathan could shew our Lord all the kingdomes of the
> world, wee are not curiously to enquire, seeing by the subtiltie of his
> wit, and knowledge gotten by long experience; hee being now almost
> six thousand yeares olde, hee can doe many things, which wee cannot
> vnderstand: for if a man by the quicknes of his wit, hath found out the
> way to present a view of the whole world in a Mappe; or to let a man
> see himselfe represented in a glasse, if this I say man can doe by the
> helpe of nature, who for knowledge, is but a childe, if he be compared
> with Angels, and for experience but of yesterday; why shall wee thinke it
> strange to heare that Sathan could make a real representation in the Aire,
> of all the glorious kingdomes of the world, which we thinke he did.[31]

The authorship and intended audiences of these texts may differ, with one
composed as a play, the other a gloss of biblical writings. Nonetheless, both
communicate a sceptical attitude towards the representational capabilities of the map and mistrusting its purpose. Where Puttock's suspicion is
enunciated explicitly, Cowper's is implied – the latter draws a direct analogy
between diabolic machination, between the 'subtiltie of [Satan's] wit' and the
'quicknes of [man's] wit'. Such wit, Cowper continues, manifests itself in two
objects: firstly, the reflective mirror or 'glasse'; and secondly, the map. Just as
Satan can distil into Jesus's view all the kingdoms of the world, so the map-
maker, with a comparable 'subtiltie' or 'wit' can contain the entire world in
a map. The widespread belief in the map's ability to surpass the immediately
visible is invoked to explicate one of the very few actions by Satan described
in the Bible. For Cowper, the world of the maps is comparable to Satanic
action in the potential malignance of its representational capacity.

The ethical duality of early modern cartographic consciousness is evident in these texts. For the author of *The Boke Named the Governour*, maps
are objects of 'pleasure' for the 'gentle wit'; while for Puttock, Daniel and
Cowper, the map is an object of suspicion, drawn into a wider schema of
ensorcellment that is powerfully paralleled with diabolic actions described
in the Bible. However, the conjectures of Puttock, Daniel and Cowper
underscore an idea which is fundamental to the prevalent contemporary perception of cartographic study and which permeated many of those scholarly

discussions which eulogised the map: the notion that the cartographic document could in some way contain a world in itself. The recurrent motifs of the map student's spatial and temporal transcendence and of closeted yet unrestricted vicarious travel serve to reinforce the extent of the prevalence of the map as a 'world on world'.

Attesting to the prevalence of the cartographic world in early modern discourse is its recurrence in a wide range of literary modes including educational treatises, biblical exegesis and poetry. In a society increasingly familiar with the map and mapping processes, there was a profound belief expressed by a wide range of thinkers in the map's unique relationship with the geography it contained, one that encompassed not only representation but also realisation. The early modern map was believed to be not only mimetic but also conjurative, evoking at once a suspicion of maleficence but also a celebration of imaginative creation.

World creation

These beliefs, exemplified by a range of encounters with the map in the study, the classroom and on the stage, importantly implicate cartography in a wider Renaissance belief in the creation of imaginative space through technological, scientific or artistic enterprise. The trend pervaded the literature of the period and is most apparent in the utopian fiction of Thomas More, Francis Bacon and Henry Neville; we witness an interest in visualised spaces, societies and cultures.[32] Such a genre appropriated world-creating for the purpose of disseminating socio-political agendas: according to Robert Appelbaum, early modern utopian fiction writers were through their writings 'actively involved in the political, religious, and scientific turmoil of their day, [with] each in [their] own way promoting progressive policies and programmes'.[33] Elsewhere, literature, in a striking parallel with the map, became its own means of creating worlds. Prose fiction, and especially romances set in quasi-fantastical lands, grew increasingly prominent and proliferated widely.[34] In Philip Sidney's *Defence of Poesie* (1595), one of the earliest works of literary theory written in the English language, world-creating is placed at the heart of literary endeavour. Sidney champions the creative capacity of 'poesie' to embellish and gild the real and also to conjure an imaginary, idealised space. '[D]isdaining to be tied to any such subjection, lifted up with the vigor of his own invention, [...] [and] freely ranging within the zodiac of his own wit', the writer, Sidney insists, 'bringeth forth, or, quite anew, forms such as never were in nature'.[35] Gavin Alexander rightly observes that this famous analogy is 'central to [Sidney's] theory of poetry',[36] and the Sidneian rhetoric of escape from the mundane ('disdaining to be tied to any such subjection') and fashioning of the writer as an imaginative traveller with a telling allusion to cosmographical formulations ('freely ranging within the zodiac of his own wit') contains echoes of the mental escapism of the map reader. Strikingly, Sidney articulates his paradigm in the language of space, referencing worlds: '[Nature's] world is brazen, the poets only deliver a golden'.[37]

The fascination with creating worlds that is so pervasive in mapping discourses permeated other branches of early modern intellectualism, demonstrating how cartography accorded with broader ideas in Renaissance thought. Elizabeth Spiller has provided an insightful examination of the 'early modern interest in the possibilities of "small worlds"' in relation to what she calls 'the larger context of Renaissance theories of worldmaking'.[38] Such a trend was part of a broader epistemological shift in European scientific thought. As Spiller writes:

> These small worlds include scientific experiments, scale models, philosophical constructs, and literary fictions. Early modern science discovered small worlds in the form of both experiments and scale models: [...] when philosophy moved away from Aristotelian emphasis on describing ordinarily accepted experience, science became a practice of discovery.[39]

Spiller's analysis of the Renaissance 'practice of discovery' also explores the alchemical experiments of Richard Eden. A scientist of seminal influence, Eden was a significant figure in the development of early modern English geographical knowledge, retaining a lifelong interest in the genesis and creation of worlds.[40] In August 1592, he carried out the common alchemical experiment of dissolving mercury and silver in nitric acid, recording the procedure and its conclusion in a letter to the Elizabethan courtier William Cecil. Upon examining the resulting compound, Eden supposed that he had somehow, through experimentation, created a new world, 'a little round iland' with its own vegetation: 'sylver trees about an ynch high [...] perfectly formed with trunks, stalkes, and leaves'.[41] This 'little round iland' was, Eden supposed, so perfect and pure that it emerged as its own unique world-like entity, for 'no lymner or paynter is able to counterfacte the like'.

The objects that Eden describes are glistening crystallised masses of mercury nitrate and silver nitrate, and the experimenter quickly conceives of the process as a method to create something ex nihilo. Eden supposes that such experiments ought to be worth 'more than a kingdom to a wise man'.[42] The resemblances with the contemporary appeal of cartography are apparent. Eden's 'little iland[s] [...] about an ynch high' are manufactured through alchemical experiments. Similarly, the map is a world created by the mapper, offering the reader cerebral nourishment, the chance to 'securely sail' and 'beholde the diuersitie of countries: natures of people, & innumerable formes of Beastes, Foules, Fishes, Trees, Frutes, Stremes, & Meatalles'. The map was part of a wider theoretical model that percolated through diverse areas of early modern thought. Typified by Sidney's understanding of the power of poetry and Eden's reflections on his scientific endeavours, this tendency postulated the theory that worlds could, through technological, scientific or artistic enterprise, be created, examined and explored.

Maps and the imagination

World-making penetrated the practice of map-making. In doing so, the cartographic world also reflected a profound belief in the power of the imagination. The paradigm of map study, which stressed the world on the page and the ability of the map reader to visualise a three-dimensional, multi-sensual space mirrored theories regarding the power of the mind's creative faculty to shape, influence and even supersede physical reality. Milton's Satan verbalises such an idea most eloquently: 'The mind is its own place, and in itself | Can make a Heaven of Hell, a Hell of Heaven' (Book 1, 254–255). Physical reality could be distorted and shaped by the creative capacity of the self, a notion which undergirds the established belief in the power of the individual imagination.

This principle was embedded in early modern intellectual culture and espoused by some of its most influential thinkers. According to Yasmin Haskell, 'It is increasingly recognized that the early modern period was an age, perhaps *the* age, of the imagination'.[43] Haskell discusses the subject of early modern diagnoses of psychological and psychosomatic disease to describe the prevailing belief in the power of the mind to create, cause and control external realities. Citing *inter alia* the work of the Flemish physician and professor of medicine at the University of Louvain Thomas Fienus, whose *De viribus imaginationis tractatus* (1608) influenced understanding of the relationship between the psychology of the pregnant mother and the physiological development of the foetus, and also Burton's *The Anatomy of Melancholy*, Haskell writes:

> [T]he spheres of influence of the imagination were considerably wider than those reserved for it by most modern psychologists, philosophers, and neuroscientists. A respectable view, if occasionally challenged, held that a diseased imagination could precipitate bodily disease not just in onself but in *others*, even remotely. The hate-filled imagination of a diseased person might be strong enough to bring about deaths in this world. By the same token, the imagination was thought capable of healing, usually by means of the emission of subtle 'vapours'. The widely acknowledged powers of witches to heal as well as harm inhabited an uneasy space between the natural and the demonic. A strong imagination could infect others with dangerous religious enthusiasms, perhaps even change the weather, but it did so by natural, not supernatural, means. Far from being the ethereal preserve of whimsical poets and artists, then, the imagination in the early modern period was invoked by radical materialists and magicians alike. [...] In short, the early modern imagination was a force to be reckoned with and a medium to be meddled with.[44]

Haskell portrays an environment teeming with theories explicitly linking the external and the internal in a concatenating dialectic of cause and effect.

The interdependency of the physical and the psychological encompassed an extensive array of subjects from demonology to religious exegesis to biology and incorporated a broad range of competing ideas about the real nature of the world and the self.

One of the most prominent figures to insist upon the potency of the creative faculty was Michel de Montaigne. In his study 'On the Power of the Imagination', the French essayist expresses his belief in the mind's capabilities, especially its capacity for influence over the external world. Sometimes translated as 'On the Force of the Imagination', it begins with a suggestive epigram drawn from medieval scholasticism, 'Fortis imaginatio generat casum [A powerful imagination generates the event]'. Elaborating on the motion of the mind as a generator of external occurrences, Montaigne writes: 'I am one of those by whom the powerful blows of the imagination are felt most strongly. Everyone is hit by it, but some are bowled over. It cuts a deep impression into me: my skill consists in avoiding it not resisting it'.[45] The declaration of the potential of the imagination is expanded further as Montaigne maintains the possibility of the psychological fluctuations manifesting physical phenomena, in particular, illness:

> [T]he sight of another man's suffering produces physical suffering in me, and my own sensitivity has often misappropriated the feelings of a third party. A persistent cougher tickles my lungs and throat. [...] The sick whom I am duty-bound to visit I visit more unwillingly than those with whom I feel less concerned and less involved. When I contemplate an illness I seize upon it and lodge it within myself: I do not find it strange that imagination should bring fevers and death to those who let it act freely and who give it encouragement.[46]

The perceived causal link between the internal and the external advanced by Montaigne is present throughout early modern European thought, justifying Haskell's characterisation of the epoch. In his 1620 edition of *Novum Organum*, Francis Bacon repeats Montaigne's central assertion of the mind's capabilities to manifest physically psychological thoughts and feelings: '[T]hough there be many things in nature which are singular and unmatched', Bacon writes, 'yet [the mind] devises for them parallels and conjugates and relatives which do not exist'.[47] The creative power of the mind is stressed – where we encounter the alien, irregular or disagreeable, our mind manufactures symmetry and regulation. In the next section, Bacon argues that 'The human understanding when it has once adopted an opinion (either as being the received opinion or as being agreeable to itself) draws all things else to support and agree with it'.[48]

The pervasiveness of the belief in the power of the imagination in early modern English culture is expressed most articulately by its greatest writer, William Shakespeare. As Jerome Mandel has shown, Shakespeare's plays frequently emphasise the fluctuating dynamic between the tangible and the imagined, the oneiric and the lucid to project a shifting landscape where the boundaries between reality and unreality are constantly shifting.[49] In *Henry V,* the playwright invokes the understanding of the mind as an agent of physical manufacture to construct on the stage a space that surpasses the circumscriptions of the theatre. In the Chorus, the audience is beseeched to think beyond 'this unworthy scaffold', 'this cock-pit' and 'this unworthy O' (Chorus, l. 1–34). To do so, they must employ their imaginative faculties:

> [S]ince a crooked figure may
> Attest in little place a million;
> And let us, ciphers to this great accompt,
> On your imaginary forces work.
> Suppose within the girdle of these walls
> Are now confined two mighty monarchies,
> Whose high upreared and abutting fronts
> The perilous narrow ocean parts asunder:
> Piece out our imperfections with your thoughts;
> Into a thousand parts divide one man,
> And make imaginary puissance;
> Think when we talk of horses, that you see them
> Printing their proud hoofs i' the receiving earth;
> For 'tis your thoughts that now must deck our kings,
> Carry them here and there; jumping o'er times,
> Turning the accomplishment of many years
> Into an hour-glass.
>
> (Chorus, 1. 16–32)[50]

'On your imaginary forces work' – in this line, the Chorus calls on the participation of the audience's creative ability. The spatial (20–24), personal (25–29) and temporal (30–32) elements of the real are to be overcome by creative action. The faith in the imagination is reiterated at the beginning of the third act. The Chorus returns to enjoin the audience's creative engagement with the action of the play:

> Thus with imagined wing our swift scene flies
> In motion of no less celerity
> Than that of thought.
>
> (Chorus, 3. 1–3)

The agency of the mind is entreated, and as a corollary, its effectiveness, or the belief in its effectiveness highlighted. The inter-scene movement is characterised as a flight of the imagination, the mind's eye of the audience speeding in this instance from the French court of the previous scene to England.[51]

The theatre of the map

There has been a considerable degree of scholarship on the theatrical representation of space within the rich literary corpus of early modern English plays. In her study, *The Cultural Geography of Early Modern Drama, 1620–1650* (2011), Julie Sanders has sought to answer questions such as 'how did early modern people think about particular kinds of habitat, space, and environment, constructed or otherwise?; [and] how were the ways they practised or inhabited these landscapes reflective of the ways in which the literary and imaginative texts of the day thought about and represented them?'[52] Other readers have analysed the similarities between dramatists and map-makers in their attempts to portray a realistic image of the world. In an important study of the interface of geography and Renaissance drama, John Gillies has traced the interchangeability of the terms 'theatre' and 'world' as encapsulated in 'the classical topos of *theatrum mundi*'.[53] Gillies has shown how the terminology of the theatrical space and the cartographic space were often interchangeable as mappers borrowed from the stage and dramatists borrowed from cartographic literature.

This linguistic interchange is most clearly demonstrated by John Speed, whose *The Theatre of the Empire of Great Britaine* (1611) was entitled and defined generically as a theatre, and the contemporary Globe Theatre, which staged plays by Shakespeare and Jonson.[54] Again, Shakespeare is perhaps the most adept early modern writer in his exploitation of the linguistic consonances of dramaturgy and cartography. In both his poetry and plays, the map is frequently invoked rhetorically as the vehicle for a metaphorical trope – alongside Thomas Kyd, Ben Jonson, George Chapman and John Fletcher, Shakespeare used the term to convey a sense of epitome or emblem.[55] In the medium of drama, Shakespeare frequently associates the geographical space with the theatrical space. In *As You Like It*, for example, the melancholic nobleman Jaques famously pronounces that 'All the world's a stage | And all the men and women merely players' (2.7.139–166). In a similar vein is Antonio's declaration in *The Merchant of Venice* that 'I hold the world but as the world Gratiano; | A stage where every man must play his part' (1.1.82–83).

Such instances exhibit Shakespeare's sensitivity to dramaturgy's straddling of both the real world and the staged world. At work is not only a playful substitution of dramatic and cartographic language. There is also an equivalence in the seemingly discrete processes of dramatic staging and map-making. Early modern intellectuals entertained a firm belief in the possibility of the world of the map and the capability of the map-maker to create a world open to travel. In the same way, Shakespeare, through the aid of the imagination of his audience, sought to construct in the enclosed space of the stage '[t]he vasty fields of France'.

The similarities between the early modern stage and the early modern map in the engagement and stimulation of the mind to transcend the real are palpable. 'Theatrical performance, like a map, an optical refraction or a mirror', David Riggs reminds us, 'reduces cosmic events to the scale of human perception'.[56] Importantly for historians of cartography, mapping is figured in the interaction between observation and reality, utilising the imagination as a means of bridging the divide between the real and the perceived. Invoking fashionable fascination with world-creating and drawing impetus from the power of the imagination to distort and even construct physical space, cartographic study represented for early modern thinkers an engagement between mind and technology, imagination and manufacture.

The creative mapper

Map-making was drawn into a broader discourse of worldmaking which permeated the early modern culture, fashioning the map-maker as a creator, much like early modern natural philosophers and writers, of novel spaces. Such was its pervasiveness, the 'world on world' of the map became a selling point for those involved in the production and publication of atlases, maps and globes. In a comment in the preface to his 1636 translation of Gerardus Mercator's *Atlas*, Henry Hexham promises the prospective reader:

> The great Monarches, Kings and Princes of this Universe, may representively in their cabinets take a view of the extention, and limits of their owne Kingdomes, and Dominions [...] Here the Noble-man and Gentleman by speculation in his closset, may travel through every Province of the whole world [...] Here the Souldier hath matter of delight minstred unto him, in beholding the place, & reading the storie, where many bloodie Battles have been fought, and many sieges performed [...] Here the Marchant sitting in his counting-house, may know what Marchandises every Countrie affordeth, what commodities it wanteth, and whither he may transport, and vent those which are most vendible, to returne gaine and profite into his purse.[57]

Hexham's blurb summarises the by now familiar benefits of map study. Monarchs, noblemen, gentlemen and merchants alike may travel through the world unhindered by expense or danger. As well as geographical traits, political, economic, social, cultural and a whole host of other features are discernible on the map. Maps, according to Hexham, allow the reader to 'view', 'travel', behold' and 'know'. The possibilities and potentialities of the cartographic document, when harnessed to the imagination, are seemingly endless. As the following chapters will demonstrate, these rich potentialities would be exploited by a whole range of writers and mappers who adopted and adapted the cartographic gaze for a multitude of purposes.

Notes

1. Thomas Elyot, *The Boke Named the Governour* (London: Thomas Berthelet, 1531), p. 35.
2. This passage from Elyot has been cited and discussed in (among others) Rhonda Lemke Sanford, *Maps and Memory in Early Modern England: A Sense of Place* (Basingstoke: Palgrave Macmillan, 2002), pp. 17–18; and Garrett A. Sullivan, *The Drama of Landscape: Land, Property, and Social Relations on the Early Modern Stage* (California: Stanford University Press, 1998), pp. 96–97.
3. David McInnis, *Mind-Travelling and Voyage Drama in Early Modern England* (Basingstoke: Palgrave Macmillan, 2012), p. 30.
4. Robert Burton, *The Anatomy of Melancholy* (Oxford: Henry Cripps, 1621), p. 351.
5. Bernhard Klein, *Maps and the Writing of Space in Early Modern England and Ireland* (Basingstoke: Palgrave Macmillan, 2001), p. 87.
6. For more about Burton's enthusiasm for cartography and his geographical knowledge, see Anne S. Chapple, 'Robert Burton's Geography of Melancholy', *Studies in English Literature 1500–1900*, 33:1 (1993), pp. 99–130. Identifying Burton as 'one of the major [map] collectors of his day', Chapple locates the *Anatomy of Melancholy* within the wider burgeoning intellectual and technological discourses of cartography and geography: 'Burton', Chapple observes, 'was embarking on his exploration of melancholia at a time when the image of the world on maps was changing with unprecedented rapidity under the pressure of new geographical discoveries; the surprising connections between these two pursuits for Burton is fruitful ground for exploration' (p. 100).
7. J.N.L. Baker, for example, has flagged up Cuningham's role, in conjunction with his fellow Cambridge geographer John Dee, in the 'develop[ment of] the mathematical side of geography as an aid to navigation'. Baker, 'Academic Geography in the Seventeenth and Eighteenth Centuries', *The Scottish Geographical Magazine*, 51:3 (1955), p. 130.
8. Bernhard Klein, 'Partial Views: Shakespeare and the Map of Ireland', *EMLS*, 4.2:3 (1998), para. 5.
9. William Cuningham, *The Cosmographical Glasse* (London: John Daye, 1559), frontispiece. For a more extensive analysis of the pictorial symbolism of the frontispiece of Cuningham's text, see S.K. Heninger, *The Cosmographical Glasse: Renaissance Diagrams of the Universe* (San Marino: Huntington Library Press, 1977), pp. 2–3.
10. Cuningham, *Cosmographical Glasse*, p. 121,
11. For a biographical overview of Norden's life, see Frank Kitchen, 'Norden, John (*c.*1547–1625)', *ODNB*, Oxford University Press, 2004; online edn. January 2008 http://www.oxforddnb.com/view/article/20250 (accessed 12th August 2014). For an exhaustive list of Norden's cartographic works, see the bibliography of Mark Netzloff's recent edition of the 1618 publication of *The Surveyor's Dialogue* (Surrey: Ashgate, 2010), pp. 214–215. For analyses of Norden's professional endeavours and his significance within English Renaissance cartographic circles, and his relationship with other significant early modern surveyors such as William Camden and Christopher Saxton, see Kitchen's essay 'John Norden (c. 1547–1625): Estate Surveyor, Topographer, County Mapmaker and Devotional Writer', *Imago Mudi*, 49 (1997), pp. 43–61. Elsewhere, J.H. Andrews' 'John Norden's Maps of Ireland', *Proceedings of the Royal Irish Academy*, 100C:5 (2000), pp. 159–206 represents an analysis both of Norden's role as a state-sponsored cartographer (Norden's Irish surveys were commissioned by the then-High Treasurer of England Robert Cecil, and

produced in 1608) and on a wider scale the role of mapping within the political context of the early seventeenth-century Jacobean plantations of Ulster. The use of maps in English colonial discourse on Ireland will be explored further in a later chapter.

12. John Norden, *The Surveyor's Dialogue (1618): A Critical Edition*, ed. Mark Netzloff (Surrey: Ashgate, 2010), p. 25.

13. *Human Geography: An Essential Anthology*, eds. John A. Agnew and David N. Livingstone (Oxford: Blackwell Publishing, 1996), pp. 2–3.

14. J.B. Harley, *The New Nature of Maps: Essays in the History of Cartography* (Baltimore: Johns Hopkins University Press, 2002), p. 53.

15. See Garret A. Sullivan, *The Drama of Landscape*, p. 239.

16. For more background on the composition of this poem, see Ann Baynes Coiro, *Robert Herrick's Hesperides and the Book Epigram Tradition* (Baltimore: Johns Hopkins University Press, 1988), p. 152.

17. All quotations are taken from *The Complete Poetry of Robert Herrick*, ed. Tom Cain and Ruth Connolly, 2 vols. (Oxford: Oxford University Press, 2013).

18. Lisa Gorton, 'John Donne's Use of Space', *EMLS*, 4:2, Special Issue 3 (1998), para. 1. See also William Empson, 'Donne the Space Man', *The Kenyon Review*, 19:3 (1957), pp. 337–399; and Julia M. Walker, 'The Visual Paradigm of "The Good-Morrow": Donne's Cosmographical Glasse', *The Review of English Studies*, 37:145 (1986), pp. 61–65.

19. All quotations are taken from John Donne, *The Major Works*, ed. John Carey (Oxford: Oxford University Press, 1990).

20. Elsewhere, Donne affords cartography a more central role in his lyrics, as for example, in 'Hymn to God my God in my Sickness' where attendant physicians are characterised as 'cosmographers' exploring the 'map' of the body of the ailing poet (7).

21. Samuel Daniel, *A Defence of Ryme* (London: R. Read, 1603), sig. G4r.

22. Bernhard Klein, 'Partial Views' (1998), para. 2.

23. *The Puritan* was included in the two Shakespearean Folios of 1664 and 1685. However, Middleton's most recent editors Gary Taylor and John Lavagnino have included the play in their authoritative two-volume edition *Collected Works of Thomas Middleton* (Oxford: Oxford University Press, 2007), pp. 509–542.

24. All quotations are taken from Taylor and Lavagnino, *Collected Works of Thomas Middleton* (2007). Puttock's concluding phrase, 'but I could nere find the Counter in the Poultrie' reveals the sergeant's inability to locate a specific place on the map and as such, his cartographic illiteracy – the Counter was a debtor's prison located near a London street between Cheapside and Cornhill called the Poultry. Such illiteracy perhaps explains, alongside his antipathy towards scholars and their 'toyes' (3.3. 9–13), Puttock's guardedness in response.

25. See, for example, the opening lines of John Donne's 'The Sun Rising': 'Busy old fool, unruly sun | Why dost thou thus, | Through windows, and through curtains call on us?' (1–3).

26. William Tyndale, *The whole workes of W. Tyndall, John Frith, and Doct. Barnes* (London: John Daye, 1573), p. 156.

27. Thomas Hobbes, *Leviathan*, ed. with an introduction and notes by J.C.A. Gaskin (Oxford: Oxford University Press 2008), pp. 407–408.

28. For a selection of the many studies of magic and necromancy in early modern European culture, see Gary K. Waite, *Heresy, Magic, and Witchcraft in Early Modern Europe* (Basingstoke: Palgrave Macmillan, 2003); Charles Zika, *Exorcising Our Demons: Magic, Witchcraft and Visual Culture in Early*

Modern Europe (Boston: Brill, 2003) and Paola Zambelli, *White Magic, Black Magic in the European Renaissance* (Boston: Brill, 2007).

29. All quotations are taken from Christopher Marlowe, *Doctor Faustus and Other Plays*, eds. David Bevington and Eric Rasmussen (Oxford: Oxford University Press, 1995).

30. Luke 4: 5–7.

31. William Cowper, *Three Heavenly Treatises, Concerning Christ* (London: London: Printed by T.S. for John Budge, 1612), pp. 231–232.

32. Thomas More, *Utopia* (1516); Francis Bacon, *New Atlantis* (1627) and Henry Neville, *The Isle of Pines* (1668). All three texts appear in the anthology *Three Early Modern Utopias: Utopia, New Atlantis and The Isles of Pines*, ed. with an introduction and notes by Susan Bruce (Oxford: Oxford University Press, 2008).

33. Robert Appelbaum, 'Utopia and Utopianism', *The Oxford Handbook of English Prose 1500–1640*, ed. Andrew Hadfield (Oxford: Oxford University Press, 2013), p. 263. For introductory overviews of the role of utopian fiction in early modern political idealism, see also J.C. Davis, *Utopia and the Ideal Society: A Study of English Utopian Writing, 1516–1700* (Cambridge: Cambridge University Press, 1983), pp. 11–40; and Amy Boesky, *Founding Fictions: Utopias in Early Modern England* (Athens, Georgia: University of Georgia Press, 1996), pp. 1–22.

34. See Steve Mentz, *Romance for Sale in Early Modern England: The Rise of Prose Fiction* (Surrey: Ashgate, 2006), pp. 1–16.

35. Philip Sidney, *The Major Works*, ed. with an introduction and notes by Katherine Duncan-Jones (Oxford: Oxford University Press, 2008), p. 216.

36. Gavin Alexander, 'Sir Philip Sidney's *Arcadia*', *The Oxford Handbook of English Prose 1500–1640*, ed. Andrew Hadfield (Oxford: Oxford University Press, 2013), p. 220.

37. Sidney, *The Major Works*, p. 216.

38. Elizabeth Spiller, *Science, Reading and Renaissance Literature: The Art of Making Knowledge, 1580–1670* (Cambridge: Cambridge University Press, 2004), p. 46.

39. Spiller, *Science, Reading and Renaissance Literature*, pp. 26–27.

40. David Gwyn, 'Richard Eden: Cosmographer and Alchemist', *The Sixteenth Century Journal*, 15:1 (1984), p. 13.

41. Eden quoted in Spiller, *Science, Reading and Renaissance Literature*, p. 25.

42. Eden quoted in Spiller, *Science, Reading and Renaissance Literature*, p. 25.

43. Yasmin Haskell, 'Introduction: When Is a Disease Not a Disease?', *Diseases of the Imagination and Imaginary Disease in the Early Modern Period*, ed. Yasmin Haskell (Turnhout: Brepols, 2011), p. 5. See also Murray W. Bundy, '"Invention" and "Imagination" in the Renaissance', *The Journal of English and Germanic Philology*, 29:4 (1930), pp. 535–545; Katherine Park, 'The Imagination in Renaissance Psychology' (unpublished M. Phil. Thesis, University of London, 1974); and Penelope Murray, 'Editor's Introduction', *Imagination: A Study in the History of Ideas*, by Barry Cocking (Routledge: London, 1990), pp. vi–xvi, for discussions of the relationship between the real and the imaginary in early modern medical discourse.

44. Haskell, 'Introduction: When Is a Disease Not a Disease?', p. 5.

45. Michel de Montaigne, *The Complete Essays*, trans. M.A. Screech (London: Penguin Classics, 2003), p. 109.

46. Montaigne, *The Complete Essays*, p. 109.

47. Francis Bacon, *The Essays*, ed. with an introduction by John Pitcher (London: Penguin Classics, 1985), pp. 278–279.

48. Bacon, *The Essays*, p. 279.
49. Jerome Mandel, 'Dream and Imagination in Shakespeare', *Shakespeare Quarterly*, 24:1 (1973), pp. 61–68.
50. All quotations are taken from William Shakespeare, *The Oxford Shakespeare: The Complete Works*, ed. John Jowett *et al.*, 2nd edition (Oxford: Oxford University Press, 2005).
51. What is invoked in these choric interludes, according to Jerome Mantel, 'is a dual imaginative-creative act: to establish the inherently illusory play-world as a real world that may bespeak a revelation to the audience requires that the audience join the poet in the imaginative-creative act'. See Mandel, 'Dream and Imagination in Shakespeare', p. 65.
52. Julie Sanders, *The Cultural Geography of Early Modern Drama, 1620–1650* (Cambridge: Cambridge University Press, 2011), p. 11.
53. John Gillies, *Shakespeare and the Geography of Difference* (Cambridge: Cambridge University Press, 1994), p. 76.
54. As Gillies observes. 'The analogy appears to have worked both ways. The world was a theatre [...] [and] the theatre was a world'. See Gillies, *Shakespeare and the Geography of Difference*, p. 76.
55. An illustrative example of Shakespeare's usage in non-dramatic works can be found in Sonnet 68, which begins 'Thus is his cheek the map of days outworn'. For a wider study of cartographic terms in early modern literature, see Henry S. Turner, 'Literature and Mapping in Early Modern England, 1520–1688', *The History of Cartography, Volume Three, Part 1: Cartography in the European Renaissance*, ed. David Woodward (Chicago: Chicago University Press, 2007), pp. 412–426.
56. David Riggs, *The World of Christopher Marlowe* (New York: Henry Holt Publishing, 2004), p. 162.
57. Henry Hexham quoted in Nick de Somogyi, 'Marlowe's Maps of War', *Christopher Marlowe and English Renaissance Culture*, eds. Darryll Grantley and Peter Roberts (London: Scolar Press, 1996), p. 99.

Part I

Politics and mapping in early modern Britain

2 'Thou by thine arte dost so anatomize'

Embodying the Map in John Speed and Michael Drayton

Having examined how early modern cartographic enthusiasts conceptualised the map, I now turn to examining how maps and the processes of mapping were specifically employed in the period, especially in the context of political discourse. Primarily, politically inflected texts would utilise the narrative potential of the cartographic gaze set out in the previous chapter to advance ideological conceptualisations of national identity. An illustrative case in point of such an impulse is John Speed's *The Theatre of the Empire of Great Britaine* (1611). Described as 'an enactment of space, place, and people'[1] and identified as the 'first sustained' cartographic description of Britain,[2] Speed's volume represents one of the landmarks of early modern English mapping. Drawing together England, Wales, Scotland and Ireland, it portrays the emergent British state in a single atlas for the first time. In his preface, Speed summarises the rationale behind his enterprise:[3]

> The State of every Kingdome well managed by prudent government, seems to me to represent a Humane body, guided by the soveraignty of the *Reasonable Soule:* the Country and Land it self representing the one, the Actions and state affaires the other. Sith therefore the excellencies of the whole are but unperfectly laid open, where either of these parts is defective, our intendment is to take a view as well of the outward Body and Lineaments of the now-flourishing British Monarchy[.] [...] And here first wee will (by example of best Anatomists) propose to the view the whole Body, and Monarchy intire (as far as conveniently wee could comprise it) and after dissect and lay open the particular Members, Veines and Joints, (I meane the Shires, Riuers, Cities, and Townes).[4]

Endeavouring to follow the 'example of best Anatomists' and 'propose to the view the *whole Body*' of the empire of Britain, Speed will attempt in the following five hundred or so pages to 'dissect' and 'lay open the particular Members, Veines and Joints' that constitute his chosen subject. The drawing of a methodological parallel between cartography and anatomy is a captivating rhetorical flourish in the context of Speed's statement of an idealised nation state ('The State of every Kingdome well managed by

DOI: 10.4324/9781003200376-4

prudent government'), especially in its illumination of the cartographer's art from, as it were, the inside. Cartography is conceptualised to the reader as a multi-disciplinary, multi-valent and multi-form endeavour. For in the process of transforming the cartographer into an anatomist, one who will emulate the model of the 'best' practitioners in the subject, Speed also transforms the geographical landmass into a human patient, one that is to be 'dissect[ed] and lay[ed] open'.

The anatomy as methodology

An exploration of the history of 'anatomisation' in western European discourse from the twelfth century onwards reveals the interdisciplinarity of Speed's frame of reference. Motivated by a striving for national unification and portrayal of the perfect state, *The Theatre*'s use of anatomical language draws attention to early modern mapping's alertness to a particular investigative method that sought to understand objects as they were and how they worked. For many theorists, anatomy functioned as an explicatory as well as exploratory process – the anatomist was expected to learn and to teach, to examine and to exhibit. In paralleling his discipline to this well-developed model, Speed's cartography evinces an awareness of the fundamental notion that mapping involved more than a replication of a discrete physical object on the page and encompassed the distinctly more nuanced process of representation.

Initially, for late medieval and early Renaissance thinkers, the idea of anatomy was explicitly linked to the study of human or animal physiology. According to the *Oxford English Dictionary*, the earliest recorded usage of the word anatomy in the English language appears in John Trevisa's translation of *De Proprietatibus Rerum* (1398) by the Franciscan physician Bartholomaeus Anglicus. A compendious encyclopaedia with references to ancient Greco-Roman sources, the *Rerum* was an influential text in medieval scholarship which enjoyed 'rapid and international circulation', emerging as a 'common reference in ecclesiastical libraries'.[5] 'Anathomya', Anglicus writes, 'is a craft and a sciens to know how the members and lymes of the body ben issette and distingwed'.[6] Anglicus's description of anatomy is characterised by three features – firstly, 'Anathomya' or anatomy is a skilled endeavour ('a craft and a sciens'); secondly, the subject of anatomy is unambiguously human; and thirdly, the aim of anatomy is the advancement of knowledge of physiology ('how the members and lymes of the body ben issete and distingwed') of the human body.

However, as conceptualisations of anatomy developed, the link between anatomical procedure and physiological focus was increasingly destabilised. After Anglicus, many began to view the anatomical subject as both concrete and abstract, designating not only the specifically bodily but also incorporating a wider, more capacious subjective focus. In Bartholomew Traheron's translation of Giovanni da Vigo's medical textbook *The most excelent worckes of chirurgery*, published in 1543, under a section entitled 'an exposition of straunge termes and unknowen symples, belongynge unto the arte' the text provides a definition of the subject of anatomical investigation: '*Anotamie*[...] sygnifyeth the cuttinge up of a mans body, or some other thing'.[7]

As the consonances between Anglicus and Vigo demonstrate, the methodological approach of the anatomy was understood to indicate a singular type of inquiry. According to Walter J. Ong, anatomy was one instance of the 'spatial model' which formed a 'key to the mental world' for Renaissance thinkers. Citing, among others, Lyly's *Anatomy of Wit* and Nashe's *Anatomy of Absurditie*, Ong's outlines the implantation of medical procedures to non-medical disciplines. In figuring types of knowledge as organic entities, scholars were able to take on the role of the dissecting anatomist, investigating their subjects through cutting up and exploring its metaphorical viscera.[8]

The anatomical subject: A survey

By invoking this mode, Speed's characterisation thus aligns cartography with prevailing scholarly methodologies. The stated intention to 'follow the example of best Anatomists' implicates the cartographic enterprise of the *Theatre*, and, more widely, the practice of mapping within a specific mode of intellectual endeavour that in the late sixteenth and early seventeenth century was employed across a range of scholarly disciplines. There emerged, writes Jonathan Sawday, a 'sudden and seemingly overwhelming fascination' for anatomy among intellectuals in early modern Europe.[9] Examining Renaissance culture and its predilection for adopting biological epistemologies, Sawday observes:

> 'Anatomy': the very word was a modish phrase, a guarantee of a text's modernity. John Lyly's *Euphues: the Anatomy of Wyt* (1578), Philip Stubbes's *Anatomy of Abuses* (1583), Thomas Nashe's *Anatomie of Absurditie* (1589) and of course Robert Burton's *Anatomy of Melancholy* are perhaps the most famous examples of an urge to appropriate the language of partition.[10]

Surveying the assortment of anatomies in the period substantiates Sawday's description of the term anatomy as a 'modish phrase'. Sawday's list of texts from the 1570s and 1580s that deployed 'anatomy' in their titles represents only a fraction of the large volume of self-consciously 'anatomical' works in the period.[11] The concluding appended phrase 'some other thing' in Traheron's translation of Vigo conveys the potential diversity of the anatomical subject and the broadening scope of the anatomical gaze in the humanist consciousness. In England, this potential was exploited extensively: Burton and Lyly's texts, as referred to previously, were accompanied by anatomies of death (by John More, 1596), of 'Popish Tyrannie' (by Thomas Bell, 1603), of 'the Controversed Ceremonies of the Church of England' (by John Sprint, 1618), of 'the English Nunnery at Lisbon' (by Thomas Robinson, 1622), of 'a Woman's Tongue' (anonymous, 1638) and 'Lieutenant Col. John Lilburne's Spirit' (by Henry Dennem 1649) among an assortment of other topics.

This diversity of subject occasionally manifested a diversity of approach, a fact which may discerned in Speed's own self-styled 'anatomy'. In the most famed anatomy of the age, Burton's study of melancholy, the author describes his method and objective:

> The anatomy of melancholy what it is. With all the kindes, causes, symptomes, prognostickes, and severall cures of it. In three maine partitions with their severall sections, members, and subsections. Philosophically, medicinally, historically, opened and cut up. By Democritus Iunior. With a satyricall preface, conducing to the following discourse.[12]

What Sawday describes as 'the language of partition' is palpable both in the texts intended approach ('Philosophically, medicinally, historically, opened and cut up') but also in its very structure. Burton frames his text around three main 'partitions', relating firstly the symptoms of melancholy, secondly the prognoses for melancholy and thirdly the cures for melancholy. In doing so, the author consciously divides his text as he does the malady, exhibiting its components as the 'partitions' serve as both a means of categorisation for the purpose of the treatise; and also a means of parting, isolating and analysing melancholy's constituent parts. A similar example to Burton's definition is found in *The Anatomie of the Minde* (1576) by Thomas Rogers:

> In consideration whereof, I have named the whole, the *Anatomie* of the mind, because the minde in them is divided, and everie parte of eyther of them sufficientlye manifested, and illustrated with many examples of Heathen men, to the bettering I hope of dissembling Christians.[13]

Rogers' approach has much in common with Burton – the subject in *The Anatomie of the Minde* is divided, knowledge of its constituents are gained, 'sufficientlye manifested, and illustrated', and the study concludes with a more complete understanding of the whole.

Speed's anatomising: A divergent approach

Correlations between the purported methodologies of Burton, Rogers and Speed are noticeable, underlining the concurrences between the respective methods of the anatomist and the cartographer. Just as Speed attempts to delineate the greater whole ('Great Britaine') through the exhibition of its constituent parts ('Shires, Riuers, Cities, and Townes'), so Burton attempts to delineate 'melancholy' through exhibiting its 'kindes, causes, symptomes, prognostickes, and severall cures'. All three writers conform to what one scholar has called the anatomy's 'synecdochic logic', a model that holds that 'a complete knowledge of parts means a complete knowledge of the whole and vice versa'.[14]

Nonetheless, Speed goes about his anatomy in a different, almost contradictory way from his fellow anatomists Rogers and Burton. Burton's 'anatomization' is in many ways a medically centred anatomy, employing a dissective method – 'the anatomy of melancholy what it is. [...] Philosophically, medicinally, historically, opened and cut up'. For the author of *The Anatomy of Melancholy*, the anatomy is an exegetical exercise that commences with an immediate dissection of melancholy into its constituents, proceeds to an investigation of the resultant parts and then concludes by reconstructing the whole, with the knowledge heretofore gained assisting in a greater understanding of the subject. Speed's anatomy, by comparison, observes the whole first – 'here first wee will [...] propose to the view the whole Body, and Monarchy intire' – and then commences the dissection – 'and after dissect and lay open the particular Members, Veines and Joints'.

This discrepancy should not be considered an oversight on the part of Speed but rather an indication of the expanding scope of the anatomy by the seventeenth century. *The Theatre* further exemplifies the fact that as well as becoming less fixed in subject, the methodological process of the anatomy was also open to interpretation and reworking. Significantly, the invocation of the anatomy in the context of cartographic enterprise is reiterated by the paratextual material accompanying the 1611 edition of *The Theatre*. Among the many commendatory poems added to the first publication of *The Theatre* is one composed by John Davies. Davies, a solicitor-general of Ireland, noted scholar and lawyer, was a friend of Speed's through their shared passion for antiquarianism, and lavished praise on his acquaintance's labour:

> The faire Hibernia, that Westerne Isle likewise,
> In every Member, Artire, Nerve, and Veine,
> Thou by thine Arte dost so Anatomize,
> That all may see each parcell without paine.[15]

As well as articulating the popular notion of travelling vicariously, though safely, by map ('That all may see each parcell without paine'), Davies also anticipates the anatomical analogy of Speed's preface by describing the cartographer's task as an anatomy: 'Thou [Speed] by thine Arte dost so Anatomize'. Demonstrating the contemporary purchase of the cartography-as-anatomy paradigm, Davies' laudatory lyric reinforces the anatomical inflection of Speed's putatively cartographic endeavour. Additionally, Davies' poem, with its emphasis upon the biological entity and its conflation of the geographical landmass to an organic object through allusion to 'every Member, Artire, Nerve, and Veine' adumbrates the ordering imperatives of Speed's mapping endeavour. As we shall see, utilising anatomy's language and its heritage of biological investigation also afforded the opportunity to ascribe to the land and, as a consequence, the nation state a sense of organic unity and processes.

Why anatomy?: Embodying the land

Primarily, *The Theatre*'s characterisation of cartography as anatomy demonstrates early modern cartography's receptivity to and interaction with wider intellectual trends. In appropriating 'the example of best Anatomists' Speed, the mapmaker exhibits a willingness to adopt and adapt methodologies from different, superficially discrete disciplines. Speed's introductory comments show an awareness of popular Renaissance methodologies and an enthusiasm for relating mapmaking to other areas of early modern thought. It suggests a cross-disciplinary sharing of language, an interchange in which both cartography and cartographer were eager participants.

More explicitly, Speed's anatomical language exhibits cartography's impulse towards transmuting the geographical landscape into a human body. The anatomy was widely and variously applied. However, its roots lay in the medical world of dissection, and by adopting such a mode Speed also adopts – metaphorically at least – its original focus on the bodily. Concluding his anatomy/cartography analogy, Speed tellingly employs personification to describe his subject: he will 'dissect and lay open the particular Members, Veines and Joints, (I meane the Shires, Riuers, Cities, and Townes)'. The use of the rhetorical devices of epistrophe, or the repetition of similar phrases at the end of separate clauses, and parison, or matching verbal structures – the synchronicity of 'Members, Veines and Joints' with 'Shires, Riuers, Cities, and Townes' – underscores the conceptual transformation of land into the body in Speed's 'intendment'. Through intricately balanced syntax, Speed brings into direct correspondence biological constituents and geographical features, foregrounding his overarching analogy.

On the other side of this exchange is Speed's figuring of the land, or the 'Empire of Great Britaine' as a body and its geographical features as anatomical parts. The claim to follow the 'example of best Anatomists' may be to a degree problematic in its assumption of a stable methodological reference point. Yet Speed was by no means alone in the early modern period in his conceptualisation of a somatic landscape. Some 42 years before the publication of *The Theatre* the scholar Gabriel Harvey writes in a letter to his close friend Edmund Spenser that '*The Earth* you knowe, is a mightie great huge body, and consisteth of many divers, and contrarie members, & vaines, and arteries, and concavities'.[16] Similarly, in his essay 'Of Empire' Francis Bacon writes of the importance of merchants to rulers: 'they are the vena porta [gate vein to the liver], and if they flourish not, a kingdom may have good limbs, but will have empty veins, and nourish little'.[17] Most famous among such examples, perhaps, is Menenius's speech in Shakespeare's

Coriolanus (1.1.76–158), which Arthur Riss wittily describes as involving 'the belly politic'[18] which draws on biological imagery to explore the functioning of human society. While Shakespeare's fable focuses upon social structures and not the geographical landscape per se as Speed and Davies do, the underlying motivation driving the adoption of biological anatomy's lexicon is consistent – the attempt to draw from the physiological orderliness of the healthy human body a template for superficially non-physiological entities.

The personification of landscape and social structures was then a conventional rhetorical strategy in English discourse by the start of the seventeenth century. Speed's invocation of the anatomical method and, by implication, his figuring of the land as a human body carried with it a key imperative – the desire to project the image of a stable, coherent, meticulously hierarchised and functioning 'empire of Great Britaine'. In Speed's introduction, cartography assumes the anatomical approach to fulfil a carefully constructed purpose. Caterina Albano provides an analysis of Speed's interdisciplinary paralleling:

> The analogy between them only becomes possible when they are subject to a systematic method of analysis which proceeds from the whole to the particular, from the external, general outline of the body/kingdom, to its internal, specific functional parts.[19]

Albano's analysis illuminates the central repercussions of Speed's rhetorical transformation of cartography into a form of anatomy. By 'propos[ing] to the view the whole Body, and Monarchy intire', the cartographer cites the human body 'to represent the political and physical unity of the organs – or places – which constitute the state as a material and rational entity'. At the core of this impulse is a political contingency, with the projection of an ordered imperial state encompassing the Atlantic archipelago and centred upon the primacy of England. The preface fosters an 'image of natural order in which single elements are perceived both functionally and hierarchically as the individual components of an ideal harmonious unity'. In Speed's model, the unity and functionality of the respective geographies of each individual country in the whole imperial realm reflect – or, at any rate, should endeavour to reflect – the unity and functionality of the organs of a healthy human body. Through the interdisciplinary adoption of the anatomist's lexicon, Speed's cartography of the empire of 'Great Britaine' is imbued with coherence and in Albano's words, 'rationalism'.

Another key indicator of Speed's unificatory impulse lies in the title of the text. Such a technique reflected a substantial tradition in cartographically inflected works by the early modern period: for example, the Abraham Ortelius influential *Theatrum Orbis Terrarum* of 1570 gained substantial traction among Renaissance map readers, while John Norton's *Theatre of the Whole World* (1606) exemplifies the terms prevalence in English mapping discourses. The titular characteristics foreground Speed's embeddedness in the traditions of cartographic practice. More profoundly, however, the conscious adoption of the designation 'theatre' also projects an image of the volume as a textual platform of enactment. The pages of Speed's atlas are fashioned as a place wherein events, ideas and narratives can be projected at the behest authorial agency, much like the action on the theatrical stage are co-ordinated by the theatrical director. Nilanjana Mukherjee's incisive examination of British colonial maps of India alerts us to ways in which cartographic texts function as spaces of the contest between mapper, map reader and mapped.[20] Maps and similar technological apparatus, Mukherjee observes, 'act as a demiurgic force that mediates experiential reality of space and imbues it with a superficial structural value'.[21] In a similar fashion, Speed's metaphorical usage of the 'theatre' enacts a circum-scriptive discourse, demarcating its subject, medium and indeed message within its own self-contained and cohesive space.

The multi-media, multi-disciplinary theatre

The wealth of information contained in *The Theatre* highlights the attention paid to the intricate construction of a British physiognomy, both in maps and in prose. Speed's text is wide-ranging in its local detail, covering topics such as mythology, geography, economy, politics and many other subjects. It was almost certainly influenced by *Britannia* (1586), William Camden's immensely popular 'chorographical description of the most flourish-ing kingdoms, England, Scotland, and Ireland, and the islands adjoining, out of the depth of antiquity'.[22] Camden's text marks a signpost in the increasingly prominent discourse of British identity, which was emerging during the sixteenth century and would culminate in the consolidation of the monarchical structures of England and Scotland in 1603 and its after-math. The *Theatre* follows a similar generic structure to its forebear, with the section on the modern county of Westmorland proving typical. As well as including a map of the area, Speed incorporates a short chorography or description of local history and of various other features of the county, encompassing many topics from notes detailing physical dimensions (length and breadth) to enumeration of churches to descriptions of 'Romane coynes here found'.[23]

However, as Speed's work contains a prominent cartographic element, the *Theatre*'s apparently rhetorical transformation of geographical land-mass into a quasi-human entity assumes a distinctly visual quality. Where the 'humorous patrician' Menenius in Shakespeare's *Corialanus* parallels the machinations of the state with the functioning of the human body as a means of projecting the orderliness of a properly operational polity, Speed's analogy goes further and parallels the identity of the land, or rather its topography with the human body. This is shown by the ordered struc-ture of the atlas, which replicates the anatomical procedure set out by the preface. The Atlantic archipelago/patient is laid out in its entirety first, incorporating England, Wales, Scotland and Ireland alongside a narrative of 'Great Britaine under the Romans' (book 1, chapter 1). England, Wales and Scotland are delineated in a map, with Ireland pushed to the margins and even off the page (book 1, chapter 2). The cartographic vista is then scaled down to a representation of England and Wales, accompanied by a description of 'the type of the flourishing of the kingdome of England' (book 1, chapter 3). What ensues is a perambulation around the shires and counties of England and Wales, with Scotland receiving its own single, country-wide cartographic description (book 3, chapter 1), followed by maps of Ireland and its constituent provinces (book 4, chapters 1–4). The empire delineated in the *Theatre* derives its unity not only from the political, social and economic connectedness of its constituent shires, counties and countries but also from the hierarchisation of its physical features as laid out in the text's organisation of cartographies. Speed's preface initiates an approach to the landscape that considers it as a body. His atlas goes on to divide this body into its constituent parts to show how each and all ultimately fit together seamlessly through their physical and, by implication, socio-cultural propinquity.

The assertion that Speed's analogy is motivated by an overtly political intention, and accordingly attempts to portray a coherent and 'rational' delineation of his subject is validated partly by the preamble to the anatomical analogy cited in the opening part of this chapter – where the 'State' is compared with the 'Humane Body' – and also partly by the context of the *Theatre*'s production, and in particular the character of its author. A beneficiary of the patronage of the prominent Elizabethan courtier Fulke Greville, Speed was professionally associated with some of the most powerful contemporary political figures of late sixteenth- and early seventeenth-century England. He was highly esteemed by an Elizabethan court whose more prominent members included William Cecil and Francis Walsingham, both described by David Buisseret as 'great cartographic enthusiasts'.[24] In 1598 Speed presented charts to Elizabeth I, while his 1601 manuscript *A Description of the Civill Warres of England* is 'commended' to the monarch.[25] Speed's loyalty to the monarch is further evidenced by his employment by Elizabeth's successor, James I: according to his biographer Sarah Bendall, Speed 'in 1605 and 1608 [...] was paid for making maps for the king, and about 1606 he was granted a coat of arms'.[26] The *Theatre*'s commendation, replete with hyperbolic praise referencing everything from the king's puissance to his personal virtue to his patronage of the arts, emphasises the extent of Speed's monarchist inclinations:

The most high, and most potent monarch, James, of Great Britaine, France, and Ireland king; and most learned defender of the faith; inlarger and uniter of the British Empire; restorer of the British name; establisher of perpetuall peace, in church, and commonwealth, president of al princely vertuees and noble arts: John Speed, his majesties most lowly and loyal subject and servant, consecrateth these his labours, though unworthy the aspect of so high an imperiall Majestie.[27]

Furthermore, the striving for unification imbedded in the anatomical analogy is unsurprising, especially in the light of wider events and Speed's status as a government-employed, state-sanctioned cartographer. 'Invested with a wealth of historical, local, and cultural information'[28] the atlas appeared in a period when the idea of a unified and coherent British state was being moulded. Only nine years before the *Theatre*'s publication, England and Scotland had been unified under one monarch by the dynastic union for the first time. In Ireland, the colonising English administration entered a new period of expansion and consolidation, as the plantations of Ulster (1606 onwards) marked a significant move by the Jacobean government into one of the last strongholds of native habitation. Meanwhile, the Flight of the Earls in 1607 had left the upper echelons of the native society stripped of a substantial Gaelic presence, leaving in its wake 'a political vacuum in Irish society [...] to be filled by an experimental British culture'.[29] In the post-1603 period especially, under the guidance of Speed's 'inlarger and uniter of the British empire' James I, the state government sought to promote a unificatory image of the four nations of England, Scotland, Wales and Ireland.

The stated approach of Speed's text – the intention to 'propose to the view the whole Body, and Monarchy intire' – aligns the text with wider political impulses towards the unification of the archipelago under one state. The titular allusion to empire is indicative of the atlas's intentions. Through the adoption of anatomical discourse, Speed's *Theatre* is cartography as a representation of geography; and, concurrently, the politically motivated projection of an image of the new imperial nation with inherent orderliness and functionality.

Drayton's anatomy

The application of the anatomical approach and the figuring of the state as a human body was by no means unique to Speed's atlas. Alongside Shakespeare, Harvey and Bacon, the land-as-body paradigm appeared in other texts and was a feature of early modern attempts to conceptualise a coherent, functioning socio-cultural and political entity. However, perhaps the most striking cartographic example of this representative mode is contained in Michael Drayton's *Poly-Olbion, or a chorographicall Description of Great Britain*. Published in the same year as the *Theatre* (1611), *Poly-Olbion* presents a 30-book, near-15,000-line chorographical epic traversal through England and has been described by one reader as part of 'the apogee of the antiquarian-patriotic mapping impulse' of early seventeenth-century England.[30] A rich and multi-farious text, it has drawn a range of responses, most notably through recent scholars such as Todd Borlik, who explore the ways in which the poem encapsulates nascent concerns about deforestation,

Frontispiece to *Poly-Olbion* by Michael Drayton (1612). Engraved by William Hole. (Reproduced by permission of The Shakespeare Folger Library)

growing industrialisation and environmental disruption.[31] While Speed's *Theatre* pandered to James's desire for the establishment of a new systematised imperial state, Drayton's land-as-body figuring was underpinned by a more subversive aspiration to exalt the previous monarch Elizabeth I. If Speed evidences cartography's susceptibility to political intention, Drayton shows how mapping, through the interchanging of the language of physiology and geography, could be utilised in the articulation of particular political inclinations.

Speed's assumption of an anatomical methodology in the endeavour of cartography is apparent in the rhetoric of the preface. By contrast, Drayton's map-cum-anatomy takes on a more overtly imagistic character: opposite a brief opening poem, 'Upon the Frontispiece', an image illustrates a female figured dressed in such a way as to make her outline approximate the shape of Britain (England, Wales and (partially) Scotland). Engraved by William Hole, the female personification of the landscape represents one of the most striking visual images of Britain in seventeenth-century English culture. More profoundly for our exploration of how cartography was utilised in the period, it presents a literal enactment of John Davies' description of Speed's mapping: 'Thou by thine Arte dos so Anatomise'. According to Thomas Cogswell, *Poly-Olbion* represents just one example of how Drayton established for himself something of 'a cottage industry out of versifying [Raphael] Holinshed and Speed'.[32] If the poem, as Cogswell suggests, does indeed serve as a translation to verse of the *Theatre*'s maps, the anthropomorphic depiction of Britain as a body mirrors this appropriation by realising pictorially the 'anatomization' which both author and audience recognised in the atlas. 'Shires' are portrayed as 'Membres', 'Rivers' as 'Veines' and 'Cities and Townes' as 'Joints'; Scotland is a head, face and shoulders; Northumberland and Durham an adipose breast; Sussex a thickset foot. In its entirety, the frontispiece image is rich with allusive imagery – among the many allegorical tropes featured are the four historic conquerors of Britain perched on the framing archway, the symbolic sceptre of power and cornucopia of fertility held by the central feminised figure, and the stylised perspective that causes the land to rise out of the ship-strewn sea to the sky.[33] *Poly-Olbion*'s fusion of human physiognomy and topographical landscape imitates Speed's figuring of the land as a body and also reiterates a recurrent motif of cartography. We can see similarities between the frontispiece and the elaborate cordiform world map produced in 1536 by the French cartographer Oronce Finé, which frames the two hemispheres of the globe as two sides of a heart; or 'Europe as a Queen' (1570), an engraving attributed to Sebastian Münster which resembles more closely *Poly-Olbion*'s feminine monarch/land hybrid.[34]

Despite such similarities, the frontispiece nonetheless retains a striking singularity. The central figure in Drayton's introductory image is an absorbing admixture of land, body and map: the posture of the female figure deliberately mimics the 'real' outline of the land – both arms are laid across the torso to avoid any protruding elbows or unsightly peninsulas, and the right foot is turned to the side to ensure a level south coast and a regular Dorset. While the body is given foundation by the rock jutting out of the sea and inscribed with the legend 'Great Britaine', the land gives shape to the body, and the land itself is, in its turn, given substantiality by the flesh-iness of the female figure. Compounding this symbiosis of body and land is the robes that half clothe 'Albion': although too ruffled for the mimetic representation of 'proper' maps, they are recognisably cartographic in appearance, depicting places, rivers, hills, mountains, churches and even towns (a large city, possibly London, can be distinguished near the centre). A dialectic between body and land is enacted as the map is embodied and the body made map. The cartographic document wraps around the female figure and its shape is subordinated to hers, and simultaneously the female figure is wrapped around and subordinated to the shape of the land.

Feminising the land

In *Poly-Olbion*'s frontispiece, the mixture of geography, physiology, cartog-raphy and anatomy exemplifies the land as a body paradigm in an arresting fashion. The aesthetic characteristics of the representation – 'British' shape, fleshy substance – emphasise a commingling of body and land, and also, through the depiction of map-emblazoned robes, make a point of remind-ing the reader of the practice of map making. Alongside this, one of the most noticeable aspects of the image is the relationship between the femi-nised landscape and those standing in the bottom left and right. Conquerors of the 'landscape', which is the central figure in the engraving, the Anglo-Saxon Hengist and the Norman William are stationed as though sentinels at the archway that leads to the land/body of Britain. Angus Vine has insight-fully suggested that the cultural variety of these generals, supplemented by Brutus and Julius Caesar, signpost the text's miscellaneous identity: 'Britain's past, they suggest, is multiple, conflicting, and irreducible to a single chronological tradition'.[35] However, they are also in an open stance, glancing backward at the overtly sexualised, half-naked female. Suggesting both defence of the land and captivation with the female, they hint at a tension in the picture between protection and possession. This tension high-lights one of the manifestations of *Poly-Olbion*'s cartographic embodying of the land: an impulse towards not only rendering the geographic fleshly but also female. By explicitly gendering the landscape, the image draws car-tography into a discourse of sexual politics which buttressed patriarchal appropriation and ownership of geographical topographies.

In rendering Britain as a woman circumscribed by her male conquerors, *Poly-Olbion*'s frontispiece locates its cartographically inflected depiction within a wider mode of representation that, in the early modern period in England, was mainly deployed for colonised or potentially colonised spaces. A typical example of this mode of representation can be found in the one country of the emergent British Empire absent in the frontispiece to Drayton's self-proclaimed 'chorographicall description of tracts, riuers, mountains, forests, and other parts of this renowned isle of Great Britain' – Ireland. One of the most prominent instances of feminised personifications of Ireland appears in Edmund Spenser's *The Faerie Queene* (1590). According to one reader, the character of Irena, who appears in Book V of the epic represents 'Spenser's allegorical name for Ireland'.[36] The regal parallels with Drayton's Olbion queen are obvious. Moreover, in John Derricke's *Image of Irelande* (1581), the island is imagined as a 'Bride of heavenlie hewe' too good for its 'Bridegrome', the rough-visaged and wicked 'Karne':

> I mervailde in my mynde,
> and therevpon did muse:
> To see a Bride of heavenlie hewe,
> an ouglie Feere to chuse.
> This Bride it is the Soile,
> the Bridegrome is the Karne,
> With writhed glibbes like wicked Sprits,
> with visage rough and stearne.[37]

'The Bride', writes Derricke, 'is the Soile', a metaphor which anticipates *Poly-Olbion* in taking as its tenor the land and its vehicle the female body. In a similar fashion, the colonial administrator Luke Gernon observes in *A Discourse on Ireland* written in 1620:

> This nymph of Ireland, is at all poynts like a yong wenche that hath the greene sicknes for want of occupying. She is very fayre of visage, and hath a smooth skinn of tender grasse [...] Her flesh is of a softe and delicat mould of earthe, and her blew vaynes trayling through every part of her like ryvoletts [...] Her breasts are round hillocks of milk-yeelding grasse, and that so fertile, that they contend with the vallyes. And betwixt her leggs (for Ireland is full of havens), she hath an open harbor, but not much frequented. She hath had goodly tresses of hayre arboribusq' comae, but the iron mills, like a sharpe toothed combe, have notted and poled her much, and in her champion partes she hath not so much as will cover her nakedness [...] It is nowe since she was drawne out of the wombe of rebellion about sixteen yeares, by'rlady nineteen, and yet she wants a husband, she is not embraced she is not hedged and diched, there is noo quicksett putt into her.[38]

The overt lecherousness of Gernon's description – imagining Ireland as a young female, 'a wenche [whose] flesh is of a softe and delicat mould of earthe' and referring to her 'open harbor, but not much frequented' – exemplifies the prominence of personification, and in particular female personification in early modern English colonial discourse. Derricke and Gernon are emblematic of a wider trend in the literature of the period – many colonies or lands to be colonised were frequently imagined as female bodies. Beyond Ireland, the discourse of colonisation formulated in gendered language extended to the New World. Famously, Walter Raleigh was to write following his voyage to South America that 'Guiana is a country that hath yet her maidenhead'.[39] Noting that '[T]opographic images of women [feature] in both the cartography and the literature of the early modern period', Rhonda Lemke Sanford observes that '[t]he land thus feminised, discovery, exploration, invasion and conquest can be figured as seduction, penetration, and rape'.[40] Gernon's description of Ireland, with its salacious observation that Ireland 'betwixt her leggs […] hath an open harbor', is suffused with this type of rhetoric. Michael Householder asserts that '[i]n the European imagination, the newly discovered Americas constituted a gendered territory of mystery that required – and even desired – exposure, penetration, and possession by European men'.[41] Michel de Certeau outlines the subtext of such a rhetorical tic:

> What really is initiated here is a colonization of the body by the discourse of power. This is writing that conquers. It will use the New World as if it were a blank, 'savage' page on which Western desire will be written. It will transform the space of the other into a field of expansion for a system of production.[42]

At play is a deliberate combination of political and sexual appropriation. To an extent, de Certeau's description of the conquering process, and its inscriptive procedure – 'it will use the New World as if it were a blank, "savage" page on which Western desire will be written' – reminds us of the female body in *Poly-Olbion*'s frontispiece, with its map-emblazoned clothing and the shaping of its limbs to mimic the outline of the land. The correspondences between Drayton's cartography and this discourse are further evidenced by the fact de Certeau's premise emerges from an analysis of Jan van der Straet's depiction of Amerigo Vespucci's discovery of America. Showing Vespucci in front of a naked female native who personifies the virgin territory of the New World, one of the most striking aspects of van der Straet's image – its inclusion of a feminine personified symbol of the land and its masculine conqueror – recurs in by *Poly-Olbion*'s frontispiece. This parallel, with its emphasis on gendered conquest, is exaggerated in the later illustration: where the feminised figure of America is shown with one conqueror, Drayton's feminised landscape is accompanied by no fewer than four.

Feminising the conqueror

However, situating *Poly-Olbion*'s map within de Certeau's theoretical model is problematic. While the text's employment of the familiar trope of land-as-female is apparent due to its inclusion of both feminised landscape and male conqueror, the land subjected to such a gendering discourse is, importantly, not a foreign country or a colony. The frontispiece does not depict Ireland as Derricke, Gernon, or Spenser do; neither is the land Guyana, imagined as a virgin by Raleigh; nor does it portray America, personified by the female native, semi-recumbent and naked in van der Straet's renowned engraving. By contrast, the subject is Britain itself, the supposed coloniser and imperial power. This discrepancy, where the 'writing that conquers' is turned upon the conqueror, and imperial desire appears displaced or even elided, suggests that de Certeau's colonialist theory is unsuitable for interpreting the image adorning Drayton's poem.

What then is the motivation of *Poly-Olbion*'s cartographic feminisation of the island of Britain? Appropriating and subjugating the female form to representations of the land, the frontispiece manifests Speed's 'anatomization', but it does not obviously share the motive identified by Sanford for such a mode of representation in colonial discourse above – 'The land thus feminised, discovery, exploration, invasion and conquest can be figured as seduction, penetration, and rape'. The rationale behind the frontispiece's personification and feminisation of the land can, I believe, be identified through a reading of *Poly-Olbion* within the context of Drayton's relationship to the structures of power in early seventeenth-century England. For, while others used the feminised landscape as a trope in descriptions of imperial and colonial endeavour, *Poly-Olbion* shows that the process of embodying and feminising the map could also be motivated by more local concerns. Drayton's text demonstrates cartography's applicability as a tool of personal expression and the possibility of map making as a means of articulating particular prejudices.

Ostensibly *Poly-Olbion* echoes the *Theatre* by imbuing the geographical landmass with a distinctly human persona to project a sense of national unity. In portraying the land as a tactile human form, it replicates the unificatory striving of Speed's text by presenting. Nonetheless, there is a striking divergence: Speed's analogy is a purely verbal one while *Poly-Olbion*'s realises the land-as-body trope visually. In doing so, it confronts the reader with one of the substantial consequences of the cartography-as-anatomy paradigm – the engrafting of the organic unity of the human upon the geographical space. Thus, the hybridised female serves as an expression of national identity and unity in a period in which the nationalities of the archipelago, and most especially Englishness, were still in flux nine years after the Union of the Crowns had installed a Scottish monarch on the throne of the kingdom.

The embodied, feminised map also serves a more particular purpose, namely the communication of Drayton's political allegiances. One of the main attributes of the central figure image is its, or rather her gender. Drayton's feminised landscape seemingly gestures towards an established convention of projecting an image of the geography of the land as a female body to vindicate colonial appropriation. However, in *Poly-Olbion*'s adoption of this idea – with the embodying gaze trained on the coloniser and not the potentially colonised – the embodied land is motivated by a desire to communicate a nostalgic longing for a feminised body politic, and in particular, Elizabeth I, who pointedly was an English-born monarch on the English throne.

The overt genderedness of the land in the frontispiece is further foregrounded by the poem's text itself: *Poly-Olbion* is replete with personifications of the land, and these personifications are overwhelmingly female. John M. Adrian, echoing Barbara C. Ewell's identification of the 'prevalence of prosopopoeia' in *Poly-Olbion*,[43] highlights the recurrence of anthropomorphic figures in descriptions of the land, outlining Drayton's 'treatment of places as fully developed "characters" in their own right'.[44] The Peak District, for example, is a 'withered Beldam' whose 'physical characteristics', Adrian writes, 'actually reflect the features of the land that she embodies'.[45] The Channel Islands of Guernsey ('Brunksey'), Jersey ('Fursey') and St. Helens ('Saint Hellens') are described as 'three mayden Iles',[46] while the 'Forrests' of Sussex ('on the Sussexian ground') are 'the daughters of the Weald'.[47]

The most substantial feminine presence in the text, outside of the frontispiece, is the figure of the nymph. Numerically, the nymphean presence in *Poly-Olbion* is substantial, totalling over a hundred. By contrast, its male equivalent – the half-man, half-beast hybrid of the satyr – appears only sixteen times, a ratio of more than ten to one in favour of the female trope. Additionally, where the satyrs are usually described with reference to their appearance, the nymphs are frequently coupled with a geographical feature: hence, the poet introduces wood nymphs,[48] sea nymphs[49] and forest nymphs.[50] On many occasions, nymphs are associated with specific places – in the argument of the first song, for example, Drayton makes mention of 'Devonshire-Nymphes',[51] while in the fourth song, the poet describes 'A troupe of stately Nymphs proud *Avon* with her brings', referring to the River Avon (p. 56).[52]

The recurrence of nymph characters, frequently associated with and occasionally personifying geographical features and locations in *Poly-Olbion* repeats the persistent theme of female embodiment of the island of Britain introduced by the frontispiece. The appearance of nymphs, figures with a rich and diverse classical heritage, emphasises Drayton's gendering impulse when representing the land. Drayton, as a poet who frequently borrowed poetic structures from ancient Greek poetry, for example, in the eclogue for his sequence *Idea: the Shepheards Garland* (1593), would have been intimately familiar with the assorted poetic utilisations of the nymph. A pervasive presence in classical literature, the nymph has a heterogeneous character. While noting that '[s]cholars face a taxonomic dilemma in discussing the female figures in Greek mythology and cult',[53] Jenifer Larson observes: 'In the odes of Pindar, written for victorious athletes from various cities, we see nymphs in a different light. [...] [S]he personifies at once the land, its familiar topographical features, and the local mythic genealogy.[54]

Larson's account of the Pindaric nymph is noteworthy because Pindar was a poet whom Drayton greatly admired. In the preamble to *Poly-Olbion*, the seventeenth century English poet describes his literary predecessor as 'unimitable' and the paradigmatic writer of 'inchanting Poemes' (p. 37). Thus, Larson's assessment that in Pindaric poetry, the nymph was deployed both as a 'represent[ation] of the city' and a 'personifi[cation] of the land' is revealing, firstly in the context of Drayton's expressed approbation for Pindar, and secondly in the wider tendency towards personification that recurs in *Poly-Olbion*'s delineations of land and landscape.

Evoking eliza

Due to the overt femininity in the text in general, readers have detected resemblances in the frontispiece to other prominent early modern female identities. On a broader level, such analyses allude to the prevalence of symbolic femininity in the pictorial culture of the period.[55] In the case of *Poly-Olbion*, several scholars have emphasised the central image's evocation of the Petrarchan mistress.[56] More importantly, for our consideration of the ways in which maps are used as means of expression, there is also a more tactile correlation to the frontispiece, namely the previous monarch Elizabeth I. This comparison is supported by the similarities between the image and Elizabeth's 'Coronation Portrait' (c. 1600–1610). In both pictures, the shape of the female form is subjugated to the aim of conveyance of a specific idea. Where the figure in the frontispiece of *Poly-Olbion* is fashioned to mimic the outline of Britain, so the young Elizabeth is cast with a certain symbolism in mind, namely the evocation of her virginal status. For example, the shoulder-length hair of Elizabeth in the portrait is symbolic of her inviolate chastity.[57]

The recurrence of this fashion for a highly stylised female body, subjected and shaped for the purposes of political, national and ideological statement, is significant for our understanding of how Drayton's text uses cartography. The image covering Drayton's text presents a nostalgic evocation of the fusion of female genderedness and monarchical power that permeated many reproductions of Elizabeth's image created during her reign. This trend, according to Albert C. Labriola, centred on 'the metaphoric and mystical identification of Elizabeth with the land [and] correlated the queen's two bodies, physical and political'.[58] Such a mode of representation was itself alert to the symbolism of cartography. Prominent examples of Elizabethan portraiture often incorporated maps or globes: in the hand-over-the-globe symbolism in George Gower's Armada portrait (1588), the sense of territorial ownership and sovereignty of the monarch is powerfully conveyed. In Marcus Gheerearts' famed Ditchley portrait, painted roughly four years later (1592), the queen is depicted against a cartographic background, shown as a sphere-transgressing colossus facing westwards, her back turned on the darkness of Catholic continental Europe (1592).

Poly-Olbion then echoes a long-established pictorial trend for fusing together body and land in symbolic representations of political sovereignty. However, it was Elizabeth's successor James who was on the throne in the year of *Poly-Olbion*'s first publication and this change at the centre of power marked a decline in Drayton's status at court. Since his arrival in London in 1590, Drayton's reputation had grown steadily, with a number of increasingly successful works flowing from his pen, culminating in the popular *Englands Heroicall Epistles* published in 1597. This growing popularity was interrupted with the death of Elizabeth in 1603. Despite an initial optimism that his aspiration to be court poet would be more firmly realised by the new monarch, the writer soon fell out of favour completely.[59]

Drayton's divergences from the prevailing political and cultural conventions are important because they point to his marginal status and, as a consequence, shed light on the motivations behind the gendering of the landscape in his most substantial literary work. Richard Helgerson, in his essay 'The Land Speaks: Cartography, Chorography, and Subversion in Renaissance England', has contended that the covering illustration to Drayton's poem represents an example of a prevailing trend towards the marginalisation of royal power in late sixteenth- and early seventeenth-century cartographic texts.[60] While acknowledging that 'The imagery of authority, the crowns and the sceptre [in *Poly-Olbion*'s frontispiece], are still monarchical', Helgerson argues that 'the monarch is now the land'.[61] However, in contrast to Helgerson's thesis that *Poly-Olbion* shifts power from the royal body to the landscape, I would argue that the overt genderedness of the frontispiece points more readily to Drayton's ambivalent attitude to prevailing Jacobean, male-led court. Thus, through the use of the embodied map, *Poly-Olbion* both harks back to the most successful period of the poet's life when female body, power and the landscape were intermingled, but also presents an image of the land that projects a divergence from the male-ruled post-1603 period.

Other aspects of *Poly-Olbion* provide an indication of Drayton's uneasy relationship with the status quo. Revealingly, the poem, which sets out to write of the wonders of 'Albion's glorious Ile' (1) never refers to Britain's newest king, even where the narrative turns to the enumeration of the monarchs of his subject – as Helgerson points out, the poem's enumeration of 'kings here sung'[62] 'conspicuously omit[s] all mention of James'.[63] Additionally, unlike the dedicatory commendation to James that introduces Speed's *Theatre*, *Poly-Olbion* is dedicated not to James but rather to his son Prince Henry: a telling feature of the text's self-fashioning, especially if we consider the role of the attraction held by the Prince of Wales ato figures marginalised by the king.[64] These twin features of the text – the omission of James in the roll call of monarchs and the dedication to the young Prince of Wales – further highlight the antipathy of Drayton to the contemporary ruler. In such a context, the Elizabethan resonances of the frontispiece are imbued with additional import, conveying *Poly-Olbion*'s nostalgic cartography. It represents part of a broader rhetorical nostalgic yearning for the Elizabethan court and, concomitantly, an aversion to the Jacobean status quo.

Speed's conceptualisation of cartography as a form of anatomy, as well as situating the practice of map making within a bourgeoning intellectual tradition, characterises cartography as a transformation of the land into a body. In doing so, it firstly shows the adaptability of cartography as a diegetic medium. Secondly, it indicates its susceptibility to politically motivated discourses as Speed attempts to express the unification of the emerging imperialist British state in the post-1603 period by appropriating for the cartographic gaze the unity of organic structure. At the same time, *Poly-Olbion* realises Speed's rhetoric imagistically. Shaping and subjugating the body to the image of the land and the practice of cartography, Drayton's frontispiece exemplifies the possibilities of the embodied map, a mode of cartographic representation which parallels the human physiology with the geography of the land. In figuring his Britain as female, Drayton first and foremost situates his text within a wider discourse which appropriates the female body as an allegorical representation of colonies and of new worlds. Yet, and in contrast to Speed, who utilises anatomical imagery in drawing the map to project a burgeoning British nation state and empire, *Poly-Olbion*'s focus is not on the British and colonial, but rather the English and the local. It evokes a time when England was ruled by a female monarch and deliberately echoes the female monarchical body and its symbolic conflation with national identity. Consequently, while the feminised personification of geographical space in Drayton's frontispiece seeks to establish the unanimity of land by adopting notions of biological unity, it also shows how cartography in the early modern period could be used as a means of expressing specific attitudes towards national identity. The fascinating image of Britain ornamenting Drayton's poem serves as a pointed allusion by the author to the Elizabethan body politic and a signal of his antipathy towards her successor.

Notes

1. Christopher Ivic, 'Mapping British Identities: Speed's *Theatre of the Empire of Great Britain*', *British Identities and English Renaissance Literature*, ed. David J. Baker and Willy Maley (Cambridge: Cambridge University Press, 2002), p. 135.
2. Andrew Gordon and Bernhard Klein, *Literature, Mapping and the Politics of Space in Early Modern England*, ed. Andrew Gordon and Bernhard Klein (Cambridge: Cambridge University Press, 2001), p. 5.
3. Speed's eminence in the late sixteenth and early seventeenth centuries is evidenced both by the nature and extent of his output and also by the frequency of his collaboration with some of the period's most renowned mapmakers. Alongside the *Theatre*, he also produced the earliest world atlas published by an Englishman entitled *A Prospect of the Famous Parts of the World* (1627). Speed collaborated with the great Netherlandish map engraver Jodocus

Hondius and was a prominent figure in a circle that included renowned cartographers such as Christopher Saxton and William Camden. For more on Speed's life, work and collaborations, see Sarah Bendall, 'Speed, John (1551/2–1629)', *ODNB* (Oxford: Oxford University Press, 2004); online edn. Jan 2008 http://www.oxforddnb.com/view/article/26093 (accessed 4th May 2021).

4. John Speed, *The Theatre of the Empire of Great Britaine* (London: William Hall, 1611), p. 1.
5. M. C. Seymour, 'Bartholomaeus Anglicus (b. before 1203, d. 1272)', *ODNB* (Oxford: Oxford University Press, 2004) http://www.oxforddnb.com/view/article/10791 (accessed 10th May 2021).
6. Bartholomaeus Anglicus, *De Proprietatibus Rerum*, trans. Johannes Trevisa (1398), I. V. xlii, p. 252.
7. Giovanni da Vigo, *The most excelent worckes of chirurgery*, trans. Bartholomew Traheron (1543), sig. ζζv/2.
8. Walter J. Ong, *Ramus: Method and the Decay of Dialogue* (Cambridge, Massachusetts: Harvard University Press, 1958), pp. 314-315.
9. Jonathan Sawday, *The Body Emblazoned: Dissection and the Human Body in Renaissance Culture* (London: Routledge, 1995), p. 44.
10. Sawday, *The Body Emblazoned*, p. 44.
11. According to Richard Sugg's survey, the conceit of the literary 'anatomy' in England in the years between 1570 and 1640 flourished. In the decade of the 1570s, three texts purporting to be 'anatomies' appeared; by the 1620s, this number had increased to 14, and by the 1640s, over twenty anatomies were published. See Richard Sugg, *Murder After Death: Literature and Anatomy in Early Modern England* (Ithaca, NY: Cornell University Press, 2007), pp. 213–215.
12. Robert Burton, *The Anatomy of Melancholy* (Oxford: Henry Cripps, 1621), title page.
13. Thomas Rodgers, *The Anatomie of the Minde* (London: John Charlewood, 1576), sig. A4r.
14. R. Grant Williams, 'Disfiguring the Body of Knowledge: Anatomical Discourse and Robert Burton's *The Anatomy of Melancholy*', *English Literary History*, 68:3 (2001), p. 596.
15. John Davies, 'To the right well deseruing Mr John Speed the Author of this worke', in Speed, *The Theatre*, sig. ¶1r.
16. Edmund Spenser and Gabriel Harvey, *Three Proper And Familiar Letters* (London: H. Bynneman, 1580), p. 12.
17. Francis Bacon, *The Essays*, ed. with an introduction by John Pitcher (London: Penguin Books, 1985), p. 119.
18. Arthur Riss, 'The Belly Politic: *Coriolanus* and the Revolt of Language', *ELH*, 59:1 (1992), p. 53. See also Nate Eastman, 'The Rumbling Belly Politic: Metaphorical Location and Metaphorical Government in *Coriolanus*', *EMLS*, 13:1 (2007).
19. Caterina Albano, 'Visible Bodies: Cartography and Anatomy', *Literature, Mapping and the Politics of Space in Early Modern England*, ed. Andrew Gordon and Bernhard Klein (Cambridge: Cambridge University Press, 2001), p. 93.
20. See Mukherjee, *Spatial Imaginings in the Age of Colonial Cartographic Reason: Maps, Landscapes, Travelogues in Britain and India* (London: Taylor and Francis, 2020).
21. Mukherjee, 'A desideratum more sublime': Imperialism's Expansive Vision and Lambton's Trigonometrical Survey of India', *Postcolonial Studies*, 14:4 (2011), p. 437.

22. William Camden, *Britannia* (London: George Latham, 1637), title page.
23. Speed, *The Theatre*, pp. 85-86.
24. David Buisseret, *The Mapmaker's Quest: Depicting New Worlds in Renaissance Europe* (Oxford: Oxford University Press, 2003), p. 65.
25. Speed, *A Description of the Civill Warres of England* (1601), single sheet.
26. Bendall, 'Speed, John (1551/2–1629)', *ODNB*.
27. Speed, *Theatre* (1611), sig. ¶1r.
28. Donald K. Smith, *The Cartographic Imagination in Early Modern England* (Surrey: Ashgate, 2008), p. 75.
29. Willy Maley, *Nation, State and Empire in English Renaissance Literature: Shakespeare to Milton* (Basingstoke: Palgrave Macmillan, 2003), p. 98.
30. Peter Barber, 'Mapmaking in England, ca. 1470–1650' The History of Cartography, Volume 2 ed. by David Woodward (Chicago: University of Chicago Press, 2007), p. 1655.
31. Todd Borlik, *Ecocriticism and Early Modern English Literature: Green Pastures* (London: Routledge, 2010), pp. 75–104.
32. Thomas Cogswell, 'The Path to Elizium "Lately Discovered": Drayton and the Early Stuart Court', *The Huntington Library Quarterly*, 54:3 (1991), p. 210.
33. The four historical figures are, clockwise from top left: Brutus of Troy, a legendary descendant of Aeneas who is described in Geoffrey of Monmouth's historical chronicle *Historia Regum Britanniae* as the founder of Britain (c. 1136); Julius Caesar, who attempted a conquest in 55BC; William the Conqueror; and Hengist, an Anglo-Saxon warrior from the fifth century AD who is purported to have founded the kingdom of Kent. For a closer examination of these figures in the frontispiece, see Margery Corbett and Ronald Lightbown, *The Comely Frontispiece: The Emblematic Title-Page in England 1550–1660* (London: Routledge, 1979), pp. 153–161.
34. For more analysis of these highly formalised types of cartographic projection, see George Kish, 'The Cosmographic Heart: Cordiform Maps of the 16th Century', *Imago Mundi*, 19 (1965), pp. 13–21, and Ruth Watson, 'Cordiform Maps since the Sixteenth Century: The Legacy of Nineteenth Century Classificatory Systems', *Imago Mundi*, 60:2 (2008), pp. 182–194.
35. Angus Vine, 'Drayton's Copious Chorography' in *Poly-Olbion: New Perspectives* ed. by Andrew McRae and Philip Schwyzer (London: Boydell & Brewer, 2020), p. 22. Sjoerd Levelt's essay 'Ordinatio and Mise-en-Page of *Poly-Olbion*' in the same volume gives an account of the production of the frontispiece and its collaborative process (p. 45), while Bernhard Klein provides a perceptive analysis of the image's function as a historical document (pp. 148–166).
36. Nicholas Canny, *Making Ireland British 1580–1650* (Oxford: Oxford University Press, 2001), p. 43. For further discussions of Spenser's use of personification in his poem, see Sheila T. Cavanagh, *Wanton Eyes and Chaste Desires: Female Sexuality in 'The Faerie Queene'* (Bloomington: Indiana University Press, 1994); Lauren Silberman, *Transforming Desire: Erotic Knowledge in Books III and IV of 'The Faerie Queene'* (Berkeley: California University Press, 1995); and Andrew Hadfield, 'Another Look at Serena and Irena', *Irish University Review*, 26:2 (1996), pp. 291–302.
37. John Derricke, *The image of Irelande with a discouerie of woodkarne* (London: John Day, 1581), sig. E3r.
38. Luke Gernon, *A Discourse on Ireland, anno. 1620* (1620), pp. 349–350. http://www.ucc.ie/celt/published/E620001/ (accessed 12th March 2013).
39. Walter Raleigh, *A Discovery of Guiana, The English Renaissance: An Anthology of Sources and Documents*, ed. Kate Aughterson (London: Routledge, 2002), p. 518.

40. Rhonda Lemke Sanford, *Maps and Memory in Early Modern England: A Sense of Place* (Basingstoke: Palgrave Macmillan, 2002), p. 54.
41. Michael Householder, 'Eden's Translations: Women and Temptation in Early America', *Huntington Library Quarterly*, 70:1 (2007), p. 14. See also Anne McClintock, *Imperial Leather: Race, Gender and Sexuality in the Colonial Contest* (New York: Routledge, 1995), p. 22.
42. Michel de Certeau, *The Writing of History*, trans. Tom Conley (New York: Columbia University Press, 1988), pp. xxv–xxvi.
43. Ewell, 'Drayton's *Poly-Olbion*', pp. 299–300.
44. John M. Adrian, *Local Negotiations of English Nationhood, 1570–1680* (Basingstoke: Palgrave Macmillan, 2011), p. 90.
45. Adrian, *Local Negotiations*, p. 90.
46. Michael Drayton, *Poly-Olbion* (London: Humphrey Lownes, 1612), p. 26.
47. Drayton, *Poly-Olbion*, p. 26.
48. Drayton, *Poly-Olbion*, pp. 8, 224.
49. Drayton, *Poly-Olbion*, p. 12.
50. Drayton, *Poly-Olbion*, p. 120.
51. Drayton, *Poly-Olbion*, p. 1.
52. Drayton, *Poly-Olbion*, p. 56.
53. Jenifer Larson, *Greek Nymphs: Myths, Culture and Lore* (Oxford: Oxford University Press, 2001), p. 3.
54. Larson, *Greek Nymphs*, p. 37.
55. There is, appropriately, a vast amount of scholarly analysis regarding representations of the female body in early modern England. For a flavour of this rich corpus, see Margaret W. Ferguson, Maureen Quilligan and Nancy J. Vickers, ed. *Rewriting the Renaissance: The Discourses of Sexual Difference in Early Modern Europe* (Chicago: University of Chicago Press, 1986); Merry E. Wiesner, *Women and Gender in Early Modern Europe* (Cambridge: Cambridge University Press, 2000); and Megan Matchinske, *Writing, Gender, and State in Early Modern England: Identity Formation and the Female Subject* (Cambridge: Cambridge University Press 2006).
56. Anne Lake Prescott, 'Drayton, Michael (1563–1631)', *ODNB* (Oxford: Oxford University Press, 2004); online edn. Jan 2008 http://www.oxforddnb.com/view/article/8042 (accessed 19th May 2021).
57. John N. King, 'Queen Elizabeth I: Representations of the Virgin Queen', *Renaissance Quarterly*, 43:1 (1990), p. 43.
58. Albert C. Labriola, 'Painting and Poetry of the Cult of Elizabeth I: The Ditchley Portrait and Donne's "Elegie: Going to Bed"', *Studies in Philology*, 93:1 (1996), p. 43.
59. Scholars have long been aware of Drayton's increasingly peripheral standing within the early seventeenth-century British court. Jean Brink has emphasised Drayton's 'outsider' status: contrasting him with the 'politically astute' Edmund Spenser, Brink argues that Drayton 'never won the favour of the powerful elite who governed England'. See Brink, *Michael Drayton Revisited* (Boston: Twayne, 1990), p. 119. Richard Hardin, while recognising Drayton's liminality, has also emphasised its positive impact on the poet's work, especially in the development of his bucolic lyrics: '[Drayton was] one of the few articulate literary spokesmen for the Country at a time when virtually all the worthwhile English poetry belongs to the ambience of Court and City'. Hardin, *Michael Drayton and the Passing of Elizabethan England* (Lawrence: University Press of Kansas, 1973), p. 136. See also Cogswell, 'The Path to Elizium "Lately Discovered"', p. 228.

60. Richard Helgerson, 'The Land Speaks: Cartography, Chorography, and Sub-version in Renaissance England', *Representations*, 16 (1986), pp. 50–85.
61. Helgerson, 'The Land Speaks', p.60.
62. Drayton, *Poly-Olbion*, pp. 279–280.
63. Helgerson, 'The Land Speaks', p. 69.
64. For a further explication of Henry's status as a focus for opposition to James, see Andrew Hadfield, 'Spenser, Drayton, and the Question of Britain', *The Review of English Studies*, 51:204 (2000), pp. 582–599.

3   Judging the plot in Spenser's
A View of the State of Ireland

In Edmund Spenser's dialogue *A View of the State of Ireland*, one of the speakers Eudoxus initiates a map-reading. Responding to his fellow dialogist Irenius and his plans for the subjugation of Ireland, Eudoxus announces, '[T]hough perhaps I am ignorant of the places, yet I will take the map of Ireland before me and make my eyes in the meanwhile my schoolmasters to guide my understanding to judge of your plot'.[1] Irenius proceeds to articulate in extensive detail his military strategy for subduing the country and its querulous inhabitants.

The significance of this moment for our understanding of cartography's utility in English Renaissance culture proceeds in part from its uniqueness: as Bruce Avery rightly observes, it is 'the only *event* in the text - the only non-verbal performance of either character'.[2] Moreover, the cartographic gaze is focalised through the lens of the political. Just as Speed's anatomy-as-cartography mode is imbued with an ideological imperative, showing mapping's adaptability to bourgeoning concepts of nationhood, so Spenser exhibits the mapper's role in colonial endeavour. The map-reading in the dialogue, signified by the multi-faceted language characteristic of Spenser, is at once an exposition of mapping's function as an aid for administrative planning and also a display of cartography's role in the process of colonisation. Moreover, the importance of the event, as indicated by its singularity, is further compounded by the context of Eudoxus's call for the map and the way in which it is used. In *A View*, a text concerned with how a persistently rebellious nation may be subdued, Spenser locates the map at the very centre of early modern colonialism, and in particular, England's ongoing and troubled attempt to resettle the country.

Maintaining the focus on the role of mapping as a means of expression of political and personal attitudes developed in the previous chapter, this discussion examines specifically the appearance of the map as an 'event' in Spenser's text. Paying close attention to Avery's assertion that the map-reading 'adumbrates [...] a movement toward greater use of maps as both tools for the expansion of empire and the fostering of a nationalistic attachment to territory',[3] this chapter explores Eudoxus's gesture in relation to specific trends of early modern English literature about Ireland, its colonisation

DOI: 10.4324/9781003200376-5

and administration. Firstly, I situate Eudoxus's map summons within the broader predisposition towards pictorial and especially cartographic depiction in colonial texts contemporaneous with *A View*. Secondly, I examine how Spenser much like Speed used the map as a canvas for the expression of a particular formulation of the country as colonisable land. Moreover, through the invocation of cartography, the writer also skilfully draws mapping into the Renaissance trope of 'emplotting', a notion which through play on the word 'plot' draws together narrative, mapped, geographical and imaginative space.

Taking the 'Mappe' of Ireland

Pointedly, Eudoxus's map is invoked in a discussion explicitly orientated motivated by the desire for establishment, implementation and exercise of English colonial power. First published in 1633, Spenser's dialogue was described by its original editor James Ware as 'intended by the author for the reformation of abuses and ill customes [in Ireland]'.[4] According to more recent editors Andrew Hadfield and Willy Maley, *A View* represents 'a key text for the interpretation of late sixteenth-century Irish cultural politics' and 'should not be ignored by students of British or colonial history'.[5] While Spenser's dialogue remained unpublished in its author's lifetime, it enjoyed significant manuscript circulation among the upper echelons of the English administration in Ireland in the 1590s: Robert Devereux, second earl of Essex and erstwhile Lord Lieutenant in Ireland, owned a copy, as did his associate the Lord Chancellor Thomas Egerton.[6] Other readers included Fynes Moryson, secretary to Charles Blount who served as Lord Deputy of Ireland under Elizabeth I, John Davies, attorney general in the country from 1603 onwards and the Cromwellian propagandist and pamphleteer John Milton.[7] Such was *A View*'s purchase among its readership, it marks, according to Patricia Coughlan, 'the founding text of modern English discourse about Ireland'.[8]

A View may be characterised as a text with an expressly colonial intent. It affords an invaluable insight into a particular aspect of sixteenth-century England's imperialist manoeuvring. The stated purpose of Spenser's dialogue is to establish a method by which Ireland, theretofore prone to frequent bursts of civil unrest, may be, in the words of Eudoxus, 'turn[ed] to good uses'.[9] In doing so, Spenser presents the reader with a work which 'must possess for scholars of colonialism and post-colonialism' and a text which represents 'arguably the most sustained and sophisticated treatment of Renaissance concepts of race and identity by a major canonical author'.[10] Of especial interest to the cartographic historian is the introduction and usage of the map in a discourse planning colonial strategy. Eudoxus's gesture occurs in the middle of a discussion vis-à-vis the establishment and administration of English military outposts: discussing with his fellow speaker Irenius how to 'reduc[e] that nation [Ireland] to better government and civility'[11] the map functions as a canvas on which Irenius will lay out for Eudoxus, purportedly 'ignorant of the places' of the country, his plans for England's forces in Ireland.[12] Therefore, the discipline of cartography is unambiguously implicated by Spenser in the ongoing English colonial project in the country. Just as Speed and Drayton would later employ the cartographic gaze to articulate a homogenised view of the disparate Atlantic archipelago, so Spenser's *View* functions within a similarly political discourse of power and imperialism.

As such, Spenser's text represents a point of interest for cartographic historians and, most especially, for those interested in mapping's role in the functioning of colonial power. Scholars of cartography such as David J. Baker, Bruce Avery and Bernhard Klein, thanks to the interdisciplinary work of J.B. Harley, have become increasingly sensitive to cartography's function in this regard. In his seminal essay 'Maps, Knowledge, and Power' Harley states emphatically that '[a]s much as guns and warships, maps have been the weapons of imperialism'.[13] By the same token, Harley asserts, '[m] aps were used to legitimize the reality of conquest and empire'.[14] This legitimisation process, according to Harley, took the following course:

> Surveyors marched alongside soldiers, initially mapping for reconnaissance, then for general information, and eventually as a tool of pacification, civilization, and exploitation in the defined colonies. But there is more to this than the drawing of boundaries for the practical political or military containment of subject populations. [...] [Maps] have been used as an aggressive complement to the rhetoric of speeches, newspapers, and written texts, or to the histories and popular songs extolling the virtues of empire.[15]

Harley's understanding of maps as 'communicators of an imperial message' derives chiefly from a conception of the map as a socially constructed artefact reflecting the intricate and multi-layered matrix of inter-personal relations surrounding its sponsorship, production and usage. Writing in collaboration with fellow cartographic historian David Woodward, Harley describes maps as 'graphic representations that facilitate a spatial understanding of things, concepts, conditions, processes, or events in the human world'.[16] Within this paradigm, maps produced by the coloniser are imbued with the concerns of colonisation. Cartography ostensibly functions as a practical endeavour to better 'see' the geography of occupied countries. Yet this apparently unbiased imperative is also aligned to a more politicised exigency as the map 'assist[s] in the maintenance of the territorial status quo' of colonial rule. When used in a colonial context, maps engage a whole series of internal dynamics encompassing coloniser, colony and colonised. In short, mappers, and especially colonial mappers, map not only to record the land but also to establish administrative power, project political control and facilitate its exercise.

This theoretical narrative may be applied to Eudoxus's map-reading. The articulated subject expressed intent and coincidences between its primary speaker Irenius and Spenser serve to highlight the fundamentally colonialist character of *A View*. Primarily, the colonialist character of the cartographic event is foregrounded by the authorial voice of *A View*. For the majority of his adult life, Spenser was engaged as an administrative official in Ireland, arriving in the country as an assistant to Lord Grey de Wilton, the Lord Deputy, in 1580. According to Christopher Highley, Spenser was representative of a 'cadre' of 'English writers, poets, intellectuals, and humanists who in the mid to late sixteenth century secured positions in the administrative and military hierarchies of colonial Ireland'.[17] By the time of *A View*'s composition (1596–1598) Spenser was enmeshed in the difficulties of the colonial enterprise in the country, his term of service encompassing numerous native rebellions, the devising and implementation of the Munster Plantation (1585 onwards) and the start of the Nine Years War (1594).

Such is the prominence of Ireland in the Spenserian biography, it is plausible to draw parallels between the author and Irenius himself, returned from Ireland with an immense knowledge of the country and armed with a detailed plan to resolve its rebelliousness. Arguments for a direct correlation between Irenius and Spenser can be supported by several descriptions of contemporary events in Ireland given by the putatively fictional speaker, descriptions which imbue the text with the piquancy of documentarian realism. For example, the graphic and extended nature of the description of the Munster famine almost certainly evidences a direct personal experience of the event.[18] Similarly, the detail of the following report from Irenius of an execution in 1577 has provoked many commentators to suggest it represents an eyewitness account on the part of Spenser himself:

> [A]t the execution of a notable traytor at Limericke, called Murrogh O-Brien, I saw an old woman, which was his foster-mother, take up his head whilst he was quartered and suck up all the blood that runne thereout, saying, that the earth was not worthy to drinke it, and therewith also steeped her face and breast, and tore her haire, crying out and shrieking most terribly.[19]

Furthermore, Irenius's begrudging admiration for Ireland's bardic culture has been seen as resonating with a poetic sensibility similar to Spenser's own literary persona. While noting Irenius's recommendation that English be established as a native language among the Irish as part of the colonising programme, Clare Carroll is careful to point out the ambivalences towards Irish linguistic and literary culture in Spenser's writing. 'Spenser's relation to the Irish language, however', writes Carroll, 'was much more ambivalent and complex than these two alternatives of love and repudiation'.[20] Carroll continues, emphasising the nuanced and often hybridised character of

Spenser's position in relation to the intricate interface of English and Gaelic in Ireland:

> Neither fluent in Irish, as the Anglo-Norman Old English who had begun the process of going Gaelic in the twelfth century were, nor merely monoglot in English, as few, if any, sixteenth-century New English bureaucrats in Ireland could have been, Spenser paid keen attention to language and had a full understanding of its power.[21]

The awareness shown by Irenius of Ireland's powerful bardic culture and prestige of its poets in the political structure of the country has been taken as a sign of Spenser's complex engagement with the issue of differentiation between the cultures of coloniser and colonised.[22] The high social status afforded the bards in Gaelic culture would have appealed to the poetically inclined Spenser. Indeed, Irenius responds to Eudoxus's accusation of an illiterate Irish populace and the diminished influence of the Irish bardic class with the retort:

> For where you say the Irish have always bin without letters, you are therein much deceived; for it is certaine, that Ireland hath had the use of letters very anciently, and long before England.[23]

Later, Irenius returns to the subject and is moved to mention the sway of the bards among Irish society. 'There is amongst the Irish', he observes, 'a certain kind of people, called Bardes, which are to them insteed poets'. Such figures are held in 'high regard', for they can both build and destroy reputations with their lyrics: 'none dare displease them', Irenius notes, 'for feare to runne into reproach through their offence, and be made infamous in the mouths of all men'. Their performances are 'taken up with a generall applause', and they are afforded 'great rewards' and 'reputation'.[24] Irenius's comments focus upon the eminence of the bardic class and suggest an underlying envy on behalf of Irenius/Spenser.[25] Irenius, as a cipher for Spenser, the colonial administrator locates the dialogue in the political machinations of early modern Ireland. Such realism, in turn, highlights the very practical motivations of the map-reading, grounding it, much like Speed's quasi-cartographic representation of the feminised island, in an active political discourse.

Mapping and conquering Ireland: A visual *view*

Spenser's map then is summoned in the context of a colonial discussion and employed in the planning of colonial enterprise. In a discussion regarding early modern Anglo-Irish cartographic relations, Bernhard Klein notes how '[T]he roles of cartographer and colonizer subtly melt into each other[.]'[26] This coalescing of mapper and administrator is exemplified in *A View* where

Irenius's programme for colonisation is imparted with the aid of the map. Harley's postulation of the map as an agent of imperial manouevring resonates through Eudoxus's call for the cartographic document to 'guide my understanding to judge of your plot'.[27] Eudoxus requires the map to facilitate a fuller comprehension of Irenius's responses, and in particular, his 'plot' for the country. Having set out an extensive history of Irish origins, related the various religious and socio-cultural characteristics of the natives – including social practices, the aforementioned description of the bardic class and pastimes (for example 'carrows' or card playing) – and surveyed the reasons why Ireland has never been completely subdued by English forces, Irenius proceeds to state his own plans for the country.[28] Prompted by Eudoxus with the query 'How then doe you think is the reformation therof [of Ireland] to be Begunne, if not by laws and ordinances', Irenius sets out his plans, and also the monetary investment and troop numbers required for his proposal.[29] Specifically, the summoning of cartographic representation is instigated by Irenius's discussion of military positions, stating his 'wish [that] the chiefe power of the army [...] be garrisoned in one country that is the strongest, and the other upon the rest that is weakest'.[30]

As well as helping to facilitate the expression of Irenius's plan for Ireland and consequently providing an insight into the utility of maps as tools of colonialism, the cartographic moment in *A View* is symbolic of a wider inclination to the visual that permeates Spenser's dialogue. In doing so, the text alerts us to the way in which visualisation – including descriptive methodologies such as cartography – performs a fundamental role in the functioning of the colonial enterprise. By its appearance, the map and the practice of cartography are infused with the desire to know through 'seeing' the native Irish.

The centrality of description in Spenser's text is revealed by its title and also its very first sentence – 'But if that country of Ireland', begins Eudoxus, 'be of so goodly and commodious a soyl, as you report'.[31] Irenius's reporting of the topography of the landforms the foundation for the dialogue's narrative. The fact that the only non-verbal event in the dialogue centres on pictorial cartographic representation stresses the inclination towards the visual. According to Julia Lupton:

> Spenser's reference in the *View* to a 'mappe of Ireland' flags a series of connections between cartography and conquest which illuminates not only the political content or argument of Spenser's tract but also its mode of discourse, its enabling fantasies of political representation. [...] The text offers [...] a view, re-view, and pre-view of Ireland's present, past and future; in each case, Spenser's work remains a survey of the land, a fundamentally geographical perspective in which the topographic, synchronically systematising, and visual ordering connotations of 'view' comprehend and organise the text's chronological moments.[32]

In emphasising the 'mode of discourse' of Spenser's *View* and its offering of 'a view, re-view, and pre-view of Ireland's present, past and future', Lupton draws attention to the focus on the visual in the text's dialogic dynamic, and its deployment by Spenser as a means to detail Irish degeneracy. Such a tendency persists throughout the dialogue, especially in the questions of Eudoxus. In response to many of his interrogations, Irenius resorts frequently to descriptive rhetoric. Evident in the first half of the dialogue, where Irenius seeks to inform Eudoxus of the country and its natives by 'outlin[ing] the abuses and cultural inferiority of the Irish',[33] the attempt to 'see' the Irish takes the form of frequent references to visual characteristics with regular allusions to the dissoluteness of the natives. Irenius's linking of physical appearance to the moral character in his portrayal is symptomatic of *A View*'s recurrent association of physical attributes to deeper moral qualities. For instance, Irenius dedicates a section of the text to describing Irish clothing and tonsorial habits: 'They [the Irish] have another custome', states Irenius, '[...] that is the wearing of Mantles, and long glibbes, which is a thicke curled bush of haire, hanging downe over their eyes, and monstrously disguising them, which are both very bad and hurtfull'.[34] The mantles, in particular, are singled out for their usefulness as outdoor garments, and in particular, their potential as outlaw clothing. Noting that 'the commoditie doth not countervaile the discommoditie', Irenius observes that the long overcoat is 'a fit house for an out-lawe, a meet bed for a rebel, and an apt cloke for a theife'.[35] The outward appearance of the Irish, and in particular their clothing habits, are taken as a fundamental signifier of their recalcitrant inclinations.

As well as expending a considerable amount of time relating the physical appearance of the Irish natives and connecting it to a perceived redundancy of native ethical character, the visual is utilised as a means of articulating the physiological degeneracy of the Irish. Irenius takes especial care to describe the emaciated bodies of the starving natives in his account of the Munster famine during the Desmond Rebellion (1569–1573):

> Out of every corner of the woods and glens they came creeping forth upon their hands, for their legs could not bear them. They looked like anatomies of death; they spake like ghosts crying out of their graves; they did eat the dead carrions, happy where they could find them; yea, and one another soon after, insomuch as the very carcasses they spared not to scrape out of their graves; and, if they found a plot of watercresses or shamrocks, there they flocked as to feast for the time, yet not able long to continue therewithal; that in short space there was almost none left, and a more populous and plentifull country suddainely left voyde of man and beast; yet sure in all that warre, there perished not many by the sort, but all by the extremitie of famine, which they themselves had wrought.[36]

In the middle of the most infamous passage of Spenser's dialogue is an arresting simile centred upon the visual appearance, intended to convey the physical decrepitude of the victims of famine – '[t]hey looked like anatomies of death'. Frequently, the articulation of appearance serves as a starting point for the development and elaboration of Irenius's views on Irish events, society, culture and politics.

The recurrent emphasis in *A View* upon the visual situates the cartographic gesture within a larger representational schema that runs through the text. It also reinforces the colonialist impetus of the dialogue by according the dialogue with well-established conventions of English commentary about Ireland. We have seen in the previous chapter how Ireland, among other geographical locations, was subjected to an embodying and gendering process by early modern English writers motivated by imperialist concerns. Running alongside such rhetoric was a prevailing inclination to devise descriptions of the country and its inhabitants in an effort to inscribe a knowable schema upon the previously alien. As we shall see, this descriptive typology with its focus on the imagistic and its obvious parallels to the cartographic purpose of creating a visual illustration of the land, was a persistent feature of English literature about Ireland in the late sixteenth and early seventeenth centuries.

Imaging Ireland: Discovering, mapping and viewing

The colonisation of Ireland had remained a hazardous enterprise for the English state for many centuries. Despite the relative success of the Anglo-Norman invasion in 1169, which established an English enclave around Dublin known as the Pale, and the passing of the Crown of Ireland Act in 1542, which designated Ireland as a kingdom subject to the English monarch Henry VIII, the internal volatility of the country in the face of English rule persisted. Geographical knowledge, or rather the lack thereof, was a cause of concern in the prolonged struggle to conquer the country. In the words of David J. Baker, even in those provinces where the English presence was most established such as Munster, the Ireland Spenser inhabited remained 'singularly vulnerable to infiltration and realignment' from an ostensibly disordered native populace as a result of the 'indeterminate cartography' of the unmapped landscape.[37] 'Since [Ireland's] co-ordinates were not fixed', Baker notes, 'they could be subjected to distorting pressures which displaced the colonisers' sense of location and direction'.[38]

This indeterminacy provoked a desire to establish an image of Ireland, especially cartographically. As the demand of the sixteenth-century diplomat Thomas Smith makes clear, the English population in Ireland were increasingly encouraged by the government in London to produce a plot or 'plat' of the country or its various regions for the better implementation of the colonial project: 'Yt is high tyme some conclusion were made', writes Smith in a letter dated January 1574 to the Lord High Treasurer William

Cecil and conveying the queen's anxieties over the drawn-out and troubled engagement of English forces on the island, 'and som plat drawen up to be followed in that enterprice in Ulster'.[39]

Writers such as Robert Payne (*A Brief Description of Ireland* (1589)), John Harington (*A Short View of the State of Ireland* (1605)) and John Davies (*A Discovery of the True Causes of Why Ireland Was Never Entirely Subdued* (1612)) are representative of the response (conscious or otherwise) to Smith's demands. Their descriptions, while verbal, evidence the desire to observe, survey and project a particular view of the country.[40] Such a spate of generically specific texts focused on projecting a picture of the country and its inhabitants typify what Patricia Palmer calls the 'relentlessly visual register' of colonial officials/writers such as Spenser.[41]

Perhaps most famous among these visually inclined representations is John Derricke's *The Image of Irelande* (1581), a text which typifies the aspiration to realise a picture of the country to better know, understand and ultimately bring it under complete English rule. Written roughly fifteen years before the composition of Spenser's dialogue, Derricke's composite of verse and the engraved image contains some of the most renowned vistas of the country and has been described as 'one of the most valuable records of Elizabethan Ireland'.[42] Derricke's *Image* claims to 'plain show' the country, characterising itself as a blazon of the Irish and Ireland. While providing the reader with a discussion of the geography of the Irish landscape in a similar fashion to Drayton's versified tour of *Poly-Olbion*, the author also presents an 'image' of the rebellious Irish native, demonstrating the text's focus on providing both a snapshot landscape and also a portrayal of its inhabitants.[43] The text contains a description of the archetypal rural Irish guerrilla soldier or 'woodkarne'. Similarly incorporated is a delineation of the native rebel. Towards the conclusion, Derricke includes a biography of the Gaelic Leinster chieftain Rory O'More ('Rorie Roge'):

> The image of Irelande with a discouerie of vvoodkarne, wherin is moste liuely expressed, the nature, and qualitie of the saied wilde Irishe woodkarne, their notable aptnesse, celeritie, and pronesse to rebellion, and by waie of argumente is manifested their originall, and ofspryng, their descent and pedigree: also their habite and apparell, is there plainly showne. The execrable life, and miserable death of Rorie Roge, that famous archtraitour to God and the croune (otherwise called Rorie Oge) is like wise discribed.[44]

As with Spenser's *View*, Derricke's *Image* establishes its intention to represent the country and its inhabitants. Prose, poetry and image are appropriated for the express purpose of visual description. The sense of striving to achieve a visualisation of the Irish landscape and native subjects is palpable: pointedly, the text is categorised by its author as an 'image', while the 'wild Irishe woodkarne' and his 'habite and apparell' are 'showne' and the 'famous archtraitour to God' O'More is 'plainly showne'.[45]

Derricke's text presents an emblematic instance of the attempt by contemporary English writers to produce an image of the Irish. The parallels between Derricke and Spenser's respective texts, most especially in their desire to portray their subject, underscore the conscious implementation of a visual mode in the propagation of the colonial enterprise: '[L]ike Derricke's *Image of Irelande*, Spenser's *View* presents arguments for the destruction of Irish Culture', writes Maryclaire Moroney, 'and with it, the destruction of recalcitrant elements of the Irishry, which position English civility and Irish savagery within the framework of a much larger religious and military struggle, one which England is divinely ordained to win'.[46] To create this dichotomy of civility and savagery between coloniser and colonised, writers first had to engineer an image of the native and link its most striking features with immoral conduct. While Spenser chooses to focus on the appurtenances of the Irish such as their hair and clothing and their potential for criminal usage, Derricke's text formulates this discourse through the use of distinction and juxtaposition, providing illustrations that contrast the disciplined English regiments with the disordered rabble of indigenous military forces.

In seeking to visualise Ireland, writers such as Derricke and Spenser continued a tradition that can be traced back to the Middle Ages. One of the most famous medieval descriptions of Ireland and Irish culture was produced by the medieval travel writer Gerald of Wales. Gerald accompanied Prince John, then Lord of Ireland and soon to be King of England, on his 1185 expedition to Ireland, an expedition intent of the consolidation on the earlier incursion into the country in the 1160s.[47] The result of this journey was the composition of *Topographia Hibernica*, a text which proved a substantial source for *A View*.[48] Gerald's account was a significant resource for Spenser's formulation of the historical narrative of the genealogy of the Irish. The influence of Gerald's pejorative descriptions of Ireland and the Irish – backward, uncivilised, improvident and overwhelmingly a grarian – among Spenser and his contemporaries cannot be underestimated.[49] Tellingly, *Topographia Hibernica* or 'The Topography of Ireland' is interested in the description of Irish geography and its socio-political, demographic and environmental characteristics. In its introduction, Gerald announces: 'it occurred to me that there was one corner of the earth, Ireland, which, from its position on the furthest borders of the globe, had been neglected by others'.[50] By way of response to this absence, Gerald:

> propose[s], therefore, to take, at least, a distinct view of this remote island, both as regards its situation and character, explaining its peculiarities, so long hidden under the veil of antiquity, and searching out both the qualities and defects of almost all things which nature has produced there[.] [...] Besides this, I propose to unravel the stupendous

wonders of nature herself, to trace the descent of the various tribes from their origin, and to describe from my own knowledge the manners and customs of many men.[51]

The author endeavours to produce a fulsome image of the country, to 'explain', 'search out', 'unravel', 'trace' and describe the native populace, their traditions and origins. The intention is extensive, and its gaze panoramic.

Drawing attention to the historical importance of cartography in this visually inflected discourse of representation, Thomas O'Loughlin has recognised in an early manuscript of the *Topographia* the inclusion of a map of Ireland by an English cartographer. O'Loughlin has speculated that this map may be identified as the work of Gerald himself, or at the very least one of his associates. Citing as evidence Gerald's interest in cartography and tracing parallels between the narrative of the text and the arrangement of the map, O'Loughlin writes, 'Once assembled the evidence reveals good grounds for asserting that the map came from the immediate circle of Giraldus, that he knew it and contributed to it, and that it reflects his view of the world'.[52] The structure of the map is revealing. It adopts the medieval cartographic convention of the T-O map, which depicted the continents of Europe, Asia and Africa encircling Jerusalem by positioning east or Europe at the top. However, Gerald forgoes the tripartite structure of such projections, where the three known continents of Asia, Africa and Europe encircle Jerusalem, opting instead for a visibly hierarchised formation of Ireland at the bottom, Britain in the middle and Europe at the top.

Gerald's text shows that the inclination towards visualising Ireland and the dissemination of descriptions and other generically similar texts in English writing about Ireland accompanied the genesis of the English colonial endeavour in the country. Furthermore, the telling inclusion of the map in an early copy of the *Topographia* points not only to the perseverance of visual representations of Ireland but also to the propinquity of mapping to the historical development of colonialism in Ireland from the twelfth century onwards. It is precisely this tradition which Spenser's text continues.

Mapping Ireland

An element of this inclination towards the visual in English literature about Ireland, as the map accompanying Gerald's account suggests, was cartographic. In the sixteenth century, the English desire to map Ireland and the implicit recognition of mapping as a matter of administrative expediency grew exponentially. Spenser, as his dialogue evidences, was fully aware of such a trend: 'Marry, so ther have bin divers good plotts devised', replies Irenius to Eudoxus in his first words in the dialogue, 'and wise counsells cast alredy about reformation of that realme'.[53] Eudoxus's map-reading taps into the visualising tendency of colonial literature while specifically echoing the impulse towards visualising Ireland cartographically. Such a representational instinct was stimulated by a political requirement which was explicitly stated, with colonial overseers like Spenser expressing awareness of the value of the map and the mapper to the exercise of colonial administration. In a letter to William Cecil written in 1582, the colonel and commander of royal forces in Munster Warham St. Leger, echoing the sentiments of Thomas Smith nearly a decade earlier, complained that 'there is no man here skilful to make a map as it ought to be'.[54]

As such accounts testify, the desire for both a map of Ireland, and, more pointedly, a map of the country according with St. Leger's purposes ('as it ought to be') was a recurring topic of conversation. Furthermore, and perhaps more revealingly, the important role of cartography in the English planting project was also recognised by the Irish. Two English mappers – Robert Bartlett and John Browne – were executed by natives as they attempted to survey Ulster. Bartlett, who would be decapitated by inhabitants of Donegal, had accompanied the Lord Deputy of Ireland Lord Mountjoy, on his expedition in 1603, which brought a conclusive end to the Nine Years War and was involved in surveying Connaught for the English government before his demise.[55] This drastic response to an English cartographer by the native Irish indicates the very real sensitivity to the potency of maps both on the part of the colonising and colonised. Mindful of the power of geographic knowledge in the struggle for territory, and the usefulness of a visual perception of the land as recorded on the page, the Irish, according to John Davies' concise yet meaningful remark, 'would not have their country discovered'.[56]

In response to the lack of native cartographic skill outlined by St. Leger, cartographers from England such as Francis Jobson, Robert Lythe and the ill-fated Bartlett were employed in the attempted systematic surveying and mapping of the country. Other important early modern English mappers such as John Norden and John Speed were also engaged in producing cartographic representations of the country.[57] Furthermore, English officials in the colony were familiar with cartographers on a professional and personal basis: Henry Sidney, Lord Deputy of Ireland over three separate terms of office, engaged a surveyor from London in the late 1560s to 'take a view' of Ireland and compose a 'platte'.[58] Spenser himself was acquainted with those involved in mapping Ireland, with his associate Laurence Nowell drawing the first accurate survey of the east coast of Ireland.[59]

This increased mapping activity was engendered in part by the growing fascination with maps in English intellectual culture as set out in previous chapters. J.H. Andrews describes the process of English surveying in the colony in extensive detail, demonstrating how the success of cartographies such as Christopher Saxton's county maps of 1579 increased the enthusiasm for a cartographic knowledge of Ireland among colonial officials. However, Andrews also takes care to point out how cartography was also identified as a means to better establish and maintain English governance. In a passage reminiscent of Harley's description of maps as tools of colonialism, Andrews adumbrates the impulse towards describing the terrain of confiscated lands, setting out its underlying motives:

> [In the second half of the sixteenth century] the country's need for cartographic skill was about to reach a level of urgency unknown to earlier generations. [...] To discourage further uprisings the government resolved to confiscate the rebels' lands and redistribute them among loyal colonists brought from the west of England. The forfeited properties lay mainly in the counties of Waterford, Cork, Kerry and Limerick. They were to be settled in large estates or 'seignories', each to be made the responsibility of a gentleman 'undertaker' supported by a carefully structured hierarchy of tenant farmers, craftsmen and labourers. The enterprise that drew them all to Ireland was the famous Munster plantation. [...] Inspired by the recent progress of land surveying in England, their conception of an Irish estate was essentially quantitative. Unfortunately the domains of Desmond and his followers had never been measured except in the kind of vague and variable unit – in this case the ploughland – that contemporary Englishmen had come to regard as obsolete. Converting mediaeval to modern, ploughlands to acres, was the most revolutionary task laid on the commissioners appointed in June 1584 to take stock of the forfeited terrain. They also had to distinguish different qualities and uses of land, to make a description of the latter, the description apparently being what Burghley's correspondent [St. Leger] would have called 'lineal' as well as verbal. Such were the proceedings that one undertaker epitomised it as 'the only ground-work' for the plantation.[60]

Imaging Ireland cartographically emerged as a fundamental aspect of the English imperial project, an endeavour which sought to arrogate and classify the land's utility before establishing a complex system of political administration. The activities of English mappers would feature prominently in the plantations of Laois-Offaly (1556), Munster (1586 onwards) and Ulster (1606 onwards). As John Breen sets out, in Ireland 'cartographical [...] practices' were utilised alongside '[a]esthetic' and 'historiographical' endeavours in the 'service of a nascent colonial process of subjection'.[61] The simultaneity of mapping and colonialism supports Harley's maxim that 'surveyors marched alongside soldiers, initially mapping for reconnaissance, then for general information, and eventually as a tool of pacification, civilization, and exploitation in the defined colonies'.[62]

Spenser's mapping moment participates in a broader narrative of the creation of a systematised and officially ratified mode of representation during the English colonial project in Ireland. So concurrent was the cartographic enterprise and the implanting endeavour, certain historians have viewed them as synonymous. As William J. Smyth asserts, 'The story of the mapping of Ireland in the sixteenth and seventeenth centuries is the story of the English construction and conquest of Ireland'.[63] By providing a colonially contextualised map-reading, *A View* draws our attention to the centrality and power of maps in the implementation of imperial strategy.

The activities of Francis Jobson serve as an informative context of the interrelationship of power and cartography delineated in Spenser's text, providing a real framework of reference for *A View*'s fictive map-reading. Jobson's surveys of Ulster in the 1590s were just one of a number of 'reconnaissance surveys' intended to satisfy Elizabeth I's express request for a cartographic representation of the province.[64] Initially, the maps were sought to project an image of the native Gaelic lordships and their boundaries. However, the possessive intent of colonialism soon manifested as his surveys included signs of the future colonisation – produced some eight years before the plantation, Jobson's map of 1598 parcelled up the land into separate coloured plots, inscribed the areas with new county names and included the seal of the English monarch, located above the land and at the top of the map. This cartographic imaging emblematised English determinations, picturing an Irish landscape literally under the English crown. As Annaleigh Margey notes, 'Jobson's maps became the embodiment of early English endeavours in Ulster, with the imposition of English political structures on the province'.[65] They provided a vital platform for those colonialists to view, interpret and eventually rule the land.

The contexts of *A View* – both within the text itself but also in its correspondences to wider tendencies among the English colonising milieu to provide an image of the country – situate Eudoxus's gesture towards the map squarely within the Harleian paradigm of '[s]urveyors marching alongside soldiers'. By the 1590s, mapping endeavours and descriptive literature attended the military-led operations of expropriation, garrisoning and plantation, an established custom which is Spenser's text powerfully emphasises.

Judging Irenius's plot

Having considered the contexts of Irenius's map-reading and shown how it reflects wider links between mapping and power in the colonial context, I would now like to consider more closely how the act functions in the specific structure of the dialogue. Early modern map-readers approached the map in a number of divergent ways. As we have seen in the preceding chapters, map enthusiasts such as Thomas Elyot, John Dee and Robert Herrick utilised the map as a means of vicarious travel. In these writers, the map emerges as a multi-faceted representative device, as cartographic representation is imbued with the ability to rise above the two-dimensional physicality of its medium, and contain in Donne's evocative description of maps as 'worlds on worlds'. Combining with the perceived power of the creative faculties of the mind, the cartographic document is presented as a means of imaginative exploration and travel. In Speed and Drayton, meanwhile, the map is appropriated for the expression of political and personal aspirations, a vehicle for expressing national unity and particular agendas. In Spenser's text, however, we glimpse mapping as a discourse of power: the reader is privy to cartography's capacity to project onto an inadvertent landscape and its autochthonous inhabitants a foreign, potentially baleful schematic representation.

Scholarly interpretations have foregrounded the role of Eudoxus's map within the exercise of colonial authority in the text. Such is the prevalence of the visual and the cartographic in *A View* that John Breen has described Spenser as occupying the dual roles of 'historiographer and cartographer'.[66] Bruce Avery, citing Richard Helgerson's exploration of cartography's role as a standardiser of national and local identity in the sixteenth century, notes that 'epistemological mistrust' of eyewitness accounts of the landscape from local guides 'led to a desire for surveillance and accuracy in calculating taxes and tithes, and so prompted a desire for an accurate description of the landscape'.[67] With the colonial context, such scepticism derives from an opaque and limited perception of the landscape. Avery identifies the appearance of the map as a catalyst for a necessary movement from ambiguity to exactitude, 'forc[ing] Irenius to shift his description from a place Eudoxus has never seen, and could only see through Irenius's confused rhetoric, to one he can visualize on paper'.[68] Consequently, the production of maps such as those by Jobson, Lythe and others was motivated by a need to create a standardised, mutually understandable image, one which could be perceived by local administrators as well as superiors in Dublin and London.

This interpretation, foregrounding a tonal change towards precision in Irenius's prescription for the colonial project in Ireland, highlights the map's function in inscribing recognisable 'image' onto the previously unknown and alien landscape for the better exercise of political, military and social control. As a result, the appearance of the map is a microcosm of the English engagement with Ireland, marking a moment where 'the text adumbrates a [...] a movement toward greater use of maps as tools both for the expansion of empire and the fostering of a nationalistic attachment to territory'.[69]

Offering another view, Joanne Woolway Grenfell addresses Avery's assertion that the mapping moment represents the presentation by Eudoxus of a means to produce, from afar, 'one unified or militaristic perspective' on Ireland within the dialogic interaction.[70] Pointing to the fact that the sixteenth-century picture of Ireland was often characterised by competing viewpoints rather than by the supposedly single perspective of cartography, Grenfell reflects upon the multiplicity of cartographies of the island in early modern English representations of the country:

Whilst contemporary Irish maps clearly do show the dominance of an English perspective in their representations of Ireland, they do not always privilege the fixed, regular, and geometric ordering of space which modern commentators sometimes characterize them as doing. [...] Not only are there several maps of the country in which information from different sources is put together to provide what is obviously an incomplete or distorted picture, but there are also maps in which multiple scales and perspectives operate within the same frame.[71]

Grenfell critiques interpretations of the map's appearance as the production of concrete, precise and mutually perceptible image of Ireland. '[R]eferences to place and uses of a cartographic register', writes Grenfell, are instead connotative of an underlying unease in the colonial voice and 'are often most revealing of anxiety about the uncertainty and transience of colonial life'.[72] Her reference to the 'dominance of an English perspective' in contemporary cartographic representations of Ireland nonetheless underscore the role of the map in the projection and exercise of colonial power. The language of the text itself shows the unambiguously utilitarian aspect of Eudoxus's invocation of cartographic representation. While Eudoxus introduces the map with the intention to 'guide my understanding to judge your plot', the gesture occurs during an exchange concerning the quantification, allocation and garrisoning of troops to aid the successful execution of Irenius's proposals. The specific, seemingly impartial usage of the map as a means of geographic representation, or more accurately, the familiarisation of the unacquainted Englishman Eudoxus with the foreign, unfamiliar country better is circumvented by a clear and categorical planning of military endeavour.

As both Grenfell and Avery demonstrate, many scholars, and especially those interested in the history of the cartographic practice, have identified the location of the map's appearance in the landscape of the dialogue as a critical, perhaps even peripeteian juncture for the dialogue. In Spenser's *A View*, cartography is presented as a unique medium of specificity, a means by which the formulated plans of Irenius are endowed with the tangibility of reality and physical perception. It marks the elimination of the previous polyvocality of the text: where previously Eudoxus was a doubting interrogator, unsure as to the practicalities of Irenius's methods, the map's appearance achieves the eradication of a mythologically inflected Irish landscape in place of a more empirically based, practically focused singular view of the country.[73]

In keeping with the shift in tone signposted by cartography's appearance, Eudoxus's production of the map represents a transformative moment in the discussion about Ireland in *A View* as the text moves from a descriptive register to a prescriptive one. Furthermore, there is a marked change in subject focus from mythography and anthropological history to the pragmatic endeavour of the plantation. Imbued in the literary tradition of the dialogue was a part was a distinct interweaving of past, present and future in the colonial narrative: at stake is what one reader describes as the 'contest for the power to "plot" Ireland – define its history and geography'.[74]

The map moment chiefly highlights the parallels between the cartographer's attempt to know the land and the colonialist imperative of expropriation, ownership and cultivation, while showing how the necessity of maps of Ireland among the English derived its impetus from colonial interest. This interest also drew motivation from the contemporary understanding of cartography as a way of expressing and exercising ownership. Imaging Ireland, whether through textual description or woodcuts, helped create a legitimising discourse through verbalising links between physical appearance and moral degeneracy or cultural inferiority. On the cartographic side, a greater knowledge of the land was understood to facilitate the extension systems of economic and political compulsion and power and also underpin such expansions.

The close relationship between maps and power was a feature of cartographic texts roughly contemporaneous to Spenser's dialogue. As the titular surveyor of John Norden's *Surveyor's Dialogue* informs the farmer whose land he has come to survey, ownership of the land and knowing the land through the medium of cartography was commingled with the establishment and function of administrative control. Responding to the farmer's claim that he has 'tast[ed] of the evil that hath followed the execution of the thing, by some like unto yourself', the Surveyor states:

> There is none [...] that holdeth Land of a Lord, but he holdeth the same by some kind of Rent or service, and when he comes to take up his Land after the death of his ancestor, or upon purchase, but he doth or ought to do homage and fealty, or one of them, unto the Lord of whom he holds it: the doing whereof, how ceremonious it is, if you be a Tenant to any such land, you know, and wherein he maketh a solemn vow and oath, to be a true Tenant unto the Lord for the Land he holdeth. And sometimes the Tenant of such a tenure, is forced to be aided by his Lord for the same Land, if he be impleaded for it: now, if such a Tenant refuse to shew his estate, or to permit his land to be seen, how performeth he his oath, to be true Tenant, and to do such services as are due to the Lord?[75]

Contained within a text described by Mark Netzloff as 'intended for a general audience' and including 'the social, legal, and agricultural components of surveying',[76] the Surveyor's response reveals the understood importance of mapping to the exercise of landholding in the early modern period. Visibility is fundamental – concealment of the topography of the land, in the opinion of the Surveyor, enables the Tenant to ignore their established obligations, or 'services as are due to the Lord'.[77]

While Spenser's text initially presents to the cartographic historian an instance of mapping's role in the colonial endeavour, it also exposes how this role operates. As with Norden's Surveyor, central to Eudoxus's use of the map is the articulation and dissemination of a certain type of knowledge (in Norden, the official view of the estate, in Spenser, the official view of the colony) geared towards the successful accomplishment of a certain type of task (in Norden, estate administration, in Spenser, colonisation). In *A View*, this strategy is conveyed in the language Eudoxus employs to summon the map: '[Y]et I will take the mappe of Ireland, and lay it before me, and make mine eyes (in the meane time) my schoole-masters, to guide my understanding to judge of your plot'. Evoking a number of imbricated 'plots' – cartographic, spatial and strategic – *A View* emphasises the map's potential for narrative inscription. In its inauguration, the map-reading moment shows the author's recognition of the Renaissance notion of the world of the map as a space in dialogue with the creative imagination. While emphasising the standardising purpose of cartography as a communicative tool and aligning Spenser's text with the English colonial literature about Ireland and its prevailing tendencies, the introduction of a map to the dialogue demonstrates an awareness of cartography's ability to act as another node in the interaction between Irenius and Eudoxus, observer and planter.

Secondly, and in a related fashion, the map functions as a transmitter of a 'plot' for the colonisation of the country. If Irenius, as Julia Lupton suggests, offers Eudoxus a 'view, re-view, and pre-view of Ireland's past, present and future',[78] the map is used as a mutually perceptible space for the communication of, in particular, the last of these 'views'. It is through the cartographic document that Eudoxus will engage not only with the geography of Irenius's Ireland but also with Irenius's strategic plan for Ireland's successful colonisation, or as Eudoxus puts it Irenius's 'plot'.

David Woodward has described cartography's role in the 'changing visual idiom of the fifteenth and sixteenth centuries', arguing that 'far from being a peripheral form of illustration [it] may have been centrally and substantively important in forming a new vision of the world'.[79] The development of Renaissance cartography, he asserts, entailed that meant that 'a new [graphic] idiom was added to the old'.[80] Woodward marks out the additive nature of cartography within the overarching register of visual representation in the early modern period. The sensitivity of Spenser's dialogue to the agglomerative characteristic of the cartographic idiom is laid open in the specific passage in which Eudoxus summons the map, a section of the text which is suffused with wordplay and multiple meaning:

> I see now all your men bestowed, but what places would you set their garrisons that they might rise out most conveniently to service? and though perhaps I am ignorant of the places, yet I will take the mappe of Ireland, and lay it before me, and make mine eyes (in the mean time) my schoole-masters, to guide my understanding to judge your plot.[81]

Ostensibly, Eudoxus seeks to envision Ireland. His eyes will be his 'school-masters', teaching him as he scans the depicted landscape. Yet this visualisation is not only trained solely on a realist, external Ireland as depicted in a survey of the country, but one that incorporates the Ireland of Irenius's plot.

Tellingly, Eudoxus professes an ignorance of the geography of the landscape ('though perhaps I am ignorant of the places') to call forth the map. However, the uncertainty of 'perhaps' in this phrase suggests a degree of qualification and inadvertently serves to ironise the speaker's testimony. Eudoxus's unfamiliarity is not all it seems: earlier comments within the dialogue evince his detailed knowledge of the topography and social arrangement of the country, crucially undermining any claims to geographic illiteracy. This can be seen in particular in the discussion of those 'many other parts of Ireland, which I have heard have bin no lesse vexed with the like stormes'. Here, Eudoxus enumerates the various native insurrections, their perpetrators and localities, pointedly without recourse to a cartographic aid for either guide or instruction.[82] Detectable is a cognisance of geographical clusters in the country, especially through the rhetorical grouping of various conterminous counties in Leinster, Connaught and Ulster: for instance, Laios ('Leix'), Kilkenny and Kildare, or Westmeath, Cavan and Louth ('Lowth'). Eudoxus's express purpose of a greater understanding of Irenius's plans and an associated desire to better conceptualise Ireland with the aid of cartography is destabilised by this previously disclosed familiarity with the country. Contrary to his professed ignorance, Eudoxus is, to an extent, knowledgeable with regard to Irish geography.

Plotting the land and the colonisation

This inconsistency between the stated purpose of Eudoxus and his geographical awareness is crucial because it suggests that the function of the map in *A View* supersedes a straightforward projection of the landscape. Rather, it serves as a means of communicating a plan beyond the lay of the land. Spenser's turn to cartography assumes a multi-functional character, one which extends beyond the merely illustrative. In his attempt 'to guide my understanding to judge your plot' Eudoxus calls forth not only the plot of the map but also the plot of colonial Ireland and the plot of Irenius. Recourse to the *Oxford English Dictionary* sets out the range and diversity of the definitions of this term and draws attention to Spenser's adroit invocation of a whole series of distinct yet interrelated entities. 'Plot' can signify:

1. A fairly small piece of ground, esp. one used for a specified purpose, such as building or gardening. 2. The site or situation of a castle, town, city, etc. 3. A small portion of a surface differing in character or aspect from the remainder; a mark, patch, spot, etc.; a stain. Also in early use: a scab on a sore. 4. A map, a plan, a scheme. 5. A plan made in secret by a group of people, esp. to achieve an unlawful end; a conspiracy. 6. A design or scheme for the constitution or accomplishment of something. 7. The plan or scheme of a literary or dramatic work; the main events

of a play, novel, film, opera, etc., considered or presented as an interrelated sequence; a storyline.

("plot, n.1–7." *OED Online*. Oxford University Press, March 2022)

The majority of these definitions find their first usage in sixteenth-century texts or earlier. In their plotting, writers like Spenser were able to draw from a rich and multi-layered lexicon as the concurrence of narrative plots and spatialised plots was an especially frequent feature of early modern English literature from prose to drama to poetry. Shakespeare proves a useful reference point to such multiplicity of meaning: in *Richard III,* for example, a play closely concurrent to *A View,* the eponymous villain boasts in his opening soliloquy of his 'Plots have I laid, inductions dangerous' (1.1.32). Similarly, in *Measure for Measure* (1604), the coincidence of geographical space and narrative story is more explicitly realised as the nobleman Vincentio declares, 'The provost knows our purpose and our plot' (4.5.2).

The implication of such verbal concurrences and their tacit denotations would have powerfully exploited the heightened awareness of cartographic methodologies of representation among contemporary thinkers. In Spenser's text, the usage of the connotative term 'plot' when taken alongside its proximity to the physical object of the map has especial resonance. Read in the context of a culture of heightened mapmindedness and underscoring a focus on the visual, *A View* dexterously mingles geographical, colonial and narrative plots. As Nandini Das points out in her discussion of the attraction the 'emplotting' held for early modern writers such as Spenser's contemporary Philip Sidney:

> Repeatedly on such occasions the actual 'plot' of the grounds of Versailles or of the host's estate [.] turned suddenly into the 'plot' of a romance, the emplotment of a romance concealed yet gesturing at the plotting of alliances and oppositions, the walker transformed into the reader, the English queen confronted with the queen of the faeries. [...] Behind such [...] spatial and visual emphasis lies an entire gamut of epistemological claims made by Sidney's contemporaries. The second half of the sixteenth century saw an unprecedented rise in the development, circulation and popularity of maps.[83]

Narrative spatialisation, interwoven with the emergent cartographic consciousness and enthusiasm of the period, functioned as a communal sphere for readers and writers. In much the same way as the standardisation of cartographic representation presented to all map-readers a single, socially constructed and calibrated 'view' of the land, so plots when cited were figured as dialogic devices, performing a fundamental role in comprehension and communication.

Revealingly, in his poetry, Spenser discloses an acquaintance with the multiple meanings of the term plot, in both its noun and verb form.

According to the *Concordance to the Poems of Edmund Spenser*, the poet employs 'plot' in his lyrical oeuvre variously, including to signify a piece of land; a plan or devised course of action; and also the setting out of a surreptitious undertaking.[84] In *A View*, Spenser displays a comparable diversity in his employment of the 'plot' – appearing no less than twenty-two times, it is used as a signifier for political strategy, of a nefarious plan, and, in the aforementioned description of the Munster famine, a piece of land.[85] Similarly, Eudoxus's invocation of the plot thus intertwines possession, cartography and colonial politics. The appearance of the map allegedly serves to educate Eudoxus as to the Irish landscape. It focuses the discussion on the exactitudes of Irenius' plans of troop numbers and locations of garrisons. However, the map's appearance as a means of development from geographical ignorance to enlightenment and an emergent locational specificity is, as we have seen, deceptive.[86] It does not represent solely a clarificatory moment in Spenser's dialogue. Instead, it involves what Bruce McLeod describes as an attempt to 'reterritorialize the seemingly incoherent and the often resistant'.[87] McLeod's identification of a repeated act, 'reterritorialization', at once highlights the unoriginality of a *View*'s mapping moment – after all, Ireland had been the subject of the English cartographic gaze from the medieval period – and also points to the intended design of Irenius's gesture.

Through calling forth the plot, the speaker aims at a more textured drawing together of image and rhetoric, colonial map and colonising narrative. Bringing to attention a fluid interface of image, narrative, political strategy and imperialist cartography, it consciously conflates territory, strategy and perception. In this regard, the manifold meanings of the plot are foreshadowed by Eudoxus's announcement that he 'might make mine eyes [...] my schoole-masters'. Articulating the pedagogic role of the map, it also foregrounds a multi-layered linguistic register as well as the persistent theme of visual representation and recognition. As we have seen, 'schoole-masters' can conceivably read as both 'schoolmasters' and also 'skull masters' or eyes.[88] The linguistic coincidence of eyes and teachers serves to reiterate the central theme of visuality in the sequence and also to draw attention to the use of visual perception and, more specifically, cartographic representation within colonial discourse. In this context of burgeoning visuality, the striking invocation of the pictorial in a textual work alerts the reader to the multivalent and multimedia connotations of the plot.

Conclusion

Within what Patricia Palmer calls the 'outflow of *Views*, *Discoveries*, *Images*, *Platts*, *Anatomies*'[89] of Ireland, Spenser's mapping moment is outwardly motivated by the desire of Eudoxus to 'better' access a singular view of the colony. Yet, by his skilful use of language, Spenser exhibits the cartographic document as a place where plans, narratives and strategies may be laid out and communicated. The map at once gestures towards the role of visualisation and cartography within the processes and procedures of colonialism and also signals Eudoxus's intention to engage with his fellow dialogist within mutually comprehensible boundaries. Just as Speed attempts to consolidate a singular notion of national identity through the cartographic gaze, so Spenser utilises mapping's rich representative potential to stage the imperatives of colonialism. The cartographic document emerges, as it would for Robert Herrick half a century later when he announced to his brother the possibility to 'safely sail' by map, a space of engagement and exploration exceeding the two-dimensional page. In the multivalent language that conveys Eudoxus's motive for turning to the map – an attempt to 'judge', 'guide' and 'understand' – we are alerted to the fact that this mutual ground is more than just a means for disseminating geographic knowledge. It also represents a multi-dimensional space for Irenius to stage his colonial administration. Furthermore, the act of map-reading is imbued with colonial imperatives as this space is open to inscription and possession, encompassing image, map, view, plan and plot.

Notes

1. Edmund Spenser, *A View of the State of Ireland: From the First Printed Edition (1633)*, ed. Andrew Hadfield and Willy Maley (Oxford: Blackwell Publishing, 1997), p. 96. All quotations will be taken from this edition. In composing this chapter, I am particularly indebted to Hadfield and Maley's extensive 'Guide to Further Reading', pp. 177–182; Maley's 'Spenser and Ireland: A Select Bibliography', *Spenser Studies: A Renaissance Poetry Annual*, 9 (1991), pp. 227–242 and 'Spenser and Ireland: An Annotated Bibliography, 1986–96', *Irish University Review*, 26:2 (1996), pp. 342–353.
2. Bruce Avery, 'Mapping the Irish Other: Spenser's *A View of the Present State of Ireland*', *ELH*, 57:2 (1990), p. 263.
3. Avery, 'Mapping the Irish Other', p. 263.
4. Spenser, *A View of the State of Ireland*, p. 7. Ware's personal experience of the early modern English experience in Ireland was, like Spenser's, extensive: Graham Parry's short biography details a political career in the country as a colonial administrator from 1634 to the post-Restoration era, a period of service incorporating election as a Dublin MP (1634 and 1637), acting as a diplomat in negotiations with indigenous rebels (late-1630s onwards) and later as an emissary to the court of Charles I regarding matters in Ireland. See Parry, 'Ware, Sir James (1594–1666)', *ODNB* (Oxford: Oxford University Press, 2004); online edn. January 2008, http://www.oxforddnb.com/view/article/28729 (accessed 25th February 2021).
5. Spenser, *A View of the State of Ireland*, p. xv.
6. Christopher Highley, *Shakespeare, Spenser, and the Crisis in Ireland* (Cambridge: Cambridge University Press, 1997), pp. 116–117.
7. William J. Smyth, *Map-making, Landscapes and Memory: A Geography of Colonial and Early Modern Ireland c. 1530–1750* (Notre Dame, Indiana: University of Notre Dame Press, 2006), p. 8.
8. Patricia Coughlan, '"Cheap and common animals": The English Anatomy of Ireland in the Seventeenth Century', *Literature and the English Civil War*, eds. Thomas F. Healy and Jonathan Sawday (Cambridge: Cambridge University Press, 1990), pp. 206–207.

9. Spenser, *A View of the State of Ireland*, p. 11.
10. Spenser, *A View of the State of Ireland*, p. xvi.
11. Spenser, *A View of the State of Ireland*, p. 11.
12. Spenser, *A View of the State of Ireland*, p. 96.
13. J.B. Harley, *The New Nature of Maps: Essays in the History of Cartography* (Baltimore: Johns Hopkins University Press, 2002), p. 57.
14. Harley, *The New Nature of Maps*, p. 57.
15. Harley, *The New Nature of Maps*, pp. 57–58.
16. J.B. Harley and David Woodward, *The History of Cartography*, vol. 1 (Chicago: University of Chicago Press, 1987), p. xvi.
17. Highley, *Shakespeare, Spenser, and the Crisis in Ireland*, p. 13. Highley traces Spenser's elongated and intimate experience of Ireland, contrasting him with figures such as the courtier and poet Barnaby Googe. See Highley, *Shakespeare, Spenser, and the Crisis in Ireland*, pp. 13–14. Michael MacCarthy-Morrogh's *The Munster Plantation: English Migration to Southern Ireland 1583–1641* (Oxford: Clarendon, 1986) presents an authoritative account of the ongoing colonial endeavour of Spenser and his contemporaries.
18. *A View of the State of Ireland*, pp. 101–102.
19. *A View of the State of Ireland*, p. 66. Andrew Hadfield, noting Irenius's claim to have witnessed the execution of the Irish rebel Murrogh O'Brien in 1577, conjectures that Spenser's arrival with Grey was not his first time in Ireland, reinforcing the sense of author/speaker correspondence. See Hadfield, *Edmund Spenser's Irish Experience* (1997), pp. 17–18. See also Paul E. McLane, 'Was Spenser in Ireland in Early November 1579?', *Notes and Queries*, 204 (1959), pp. 99–101.
20. Clare Carroll, 'Spenser and the Irish Language: The Sons of Milesio in *A View of the Present State of Ireland*, *The Faerie Queene*, Book V, and the *Leabhar Gabhála*', *Irish University Review*, 26:2 (1996), p. 281.
21. Carroll, 'Spenser and the Irish Language', p. 281. For a more extensive discussion of Spenser's identity in Ireland, see Richard A. McCabe, 'Edmund Spenser, Poet of Exile', *Proceedings of the British Academy*, 80 (1991), pp. 73–103.
22. *A View of the State of Ireland*, pp. 46–47, 64, 75–77, 138.
23. Spenser, *A View of the State of Ireland*, p. 47.
24. Spenser, *A View of the State of Ireland*, p. 75.
25. As Hadfield and Maley point out, this characterisation of Irish bards provides a striking contradistinction to another poet described in the Spenserian canon, namely the fictive Bonfont, an ill-fated lyricist in Book V of *The Faerie Queene* whose tongue is nailed to a post for perceived criticisms of the Queen, Mercilla. Spenser, *A View of the State of Ireland*, pp. xx–xii.
26. Bernhard Klein, *Maps and the Writing of Space in Early Modern Britain and Ireland* (Basingstoke: Palgrave Macmillan, 2002), p. 64.
27. Spenser, *A View of the State of Ireland*, p. 96.
28. Spenser, *A View of the State of Ireland*, pp. 46–69, 75–77, 84–88.
29. Spenser, *A View of the State of Ireland*, pp. 93–96.
30. Spenser, *A View of the State of Ireland*, p. 96.
31. Spenser, *A View of the State of Ireland*, p. 11.
32. Julia Lupton, 'Mapping Mutability: Or, Spenser's Irish plot', *Representing Ireland: Literature and the Origins of Conflict 1534–1660*, eds. Brendan Bradshaw, Andrew Hadfield and Willy Maley (Cambridge: Cambridge University Press, 1993), p. 95.
33. Spenser, *A View of the State of Ireland*, p. xviii.
34. Spenser, *A View of the State of Ireland*, p. 56.

35. Spenser, *A View of the State of Ireland*, p. 57. For more on the legal underpinnings of Spenser's colonial argument and its portrayal of a native 'barbarous populace', see Brian Lockey, 'Conquest and English Legal Identity in Renaissance Ireland', *Journal of the History of Ideas*, 65:4 (2005), especially pp. 543–549. D. Alan Orr explores the continuities in the legal frameworks of Spenser's dialogue and John Davies' later similarly inclined work *A Discovery of the True Causes why Ireland was never entirely subdued* (1612). See his essay 'From a View to a Discovery: Edmund Spenser, Sir John Davies, and the Defects of Law in the Realm of Ireland', *Canadian Journal of History*, 38:3 (2003), pp. 395–408.

36. Spenser, *A View of the State of Ireland*, pp. 101–102.

37. David J. Baker, 'Off the Map: Charting Uncertainty in Renaissance Ireland', *Representing Ireland: Literature and the Origins of Conflict 1534–1660*, ed. Brendan Bradshaw, Andrew Hadfield and Willy Maley (Cambridge: Cambridge University Press, 1993), pp. 82–83.

38. Baker, 'Off the Map: Charting Uncertainty in Renaissance Ireland', p. 83.

39. Cited in Lorna Hutson, 'Fortunate Travelers: Reading for the Plot in Sixteenth-Century England', *Representations*, 41 (1993), p. 86.

40. Other English writers who sought to describe Ireland and produced accounts which conformed generically to the predominance of the visual include Fynes Moryson ('A Description of Ireland' (c. 1617)) and Peter Walsh (*A Prospect of the State of Ireland* (1682)). Perhaps most prolific of all was the poet, writer and soldier Barnabe Rich whose descriptive works on Ireland include *A Short Survey of Ireland* (1609), *A New Description of Ireland* (1610) and *Anatomy of Ireland* (1615). For more information about Rich and his military career in Ireland, see John León Lievsay, 'A Word about Barnaby Rich', *The Journal of English and Germanic Philology* (1956), pp. 381–392. For a more extensive survey of 'descriptions' of Ireland in later sixteenth-century English literature, see *Elizabethan Ireland: A Selection of Writings by Elizabethan Writers on Ireland*, ed. J.P. Myers (Hamden, Connecticut: Archon, 1983). Patricia Coughlan's '"Cheap and Common Animals": The English Anatomy of Ireland in the Seventeenth Century' (1990) and Sheila T. Cavanagh's '"The Fatal Destiny of That Land": Elizabethan Views of Ireland', pp. 116–131 provide sensitive analyses of the characteristics of this mode of colonial writing.

41. Patricia Palmer, *Language and Conquest in Early Modern Ireland: English Renaissance Literature and Elizabethan Imperial Expansion* (Cambridge: Cambridge University Press, 2001), p. 40.

42. Andrew Hadfield, 'Derricke, John (fl. 1578–1581)', *ODNB* (Oxford: Oxford University Press, 2004) online edn. January 2008, http://www.oxforddnb.com/view/article/7537 (accessed 3rd March 2021).

43. Derricke, *The Image of Irelande with a discouerie of vwoodkarne* (London: John Day, 1581), sigs. B3r–F3r.

44. Derricke, *The Image of Irelande*, sig. A1r.

45. James A. Knapp argues that Derricke's *Image* emerged 'at a crucial turning point in the course of English policy towards Ireland' (Knapp, *Illustrating the Past in Early Modern England* (Surrey: Ashgate, 2003), p. 210).

46. Maryclaire Moroney, 'Apocalypse, Ethnography, and Empire in John Derricke's *Image of Irelande* (1581) and Spenser's *View of the Present State of Ireland* (1596)', *English Literary Renaissance*, 29:2 (1999), p. 359.

47. Robert Bartlett, 'Gerald of Wales (c.1146–1220x23)', *ODNB* (Oxford: Oxford University Press, 2004), online edn. January 2008, http://www.oxforddnb.com/view/article/10769 (accessed 27th June 2014).

48. For a more extensive discussion of Spenser's borrowings from Gerald or to use his Latinised name Giraldus Cambrensis, see Hadfield, *Edmund Spenser's*

Irish Experience, pp. 28–31. 'Although not mentioned in *A View*', Hadfield writes, 'much of Spenser's analysis of Irish customs would seem to owe a great to Gerald' (p. 29). Hadfield concurs with Walter J. Ong's earlier observation that one of Spenser's key sources, Raphael Holinshed's *The Second Volume of Chronicles: conteining the Description, Conquest, Inhabitation, and Trouble- some Estate of Ireland* 'rel[ies] largely upon the very important sources, the *Topographia Hibernica* and the *Expugnutio Hibernica* of Sylvester Giraldus Cambrensis (Gerald of Wales, or Gerald de Barri)' (Ong, 'Spenser's *View* and the Tradition of the "Wild" Irish', *Modern Language Quarterly*, 3:4 (1942), p. 563). For an examination of Gerald's writings on Ireland, see F.X. Martin, 'Gerald of Wales, Norman Reporter on Ireland', *Studies: An Irish Quarterly Review* (1969), pp. 279–292 and also Andrew Murphy, *But the Irish Sea Betwixt Us: Ireland, Colonialism, and Renaissance Literature* (Kentucky: University of Kentucky Press, 1999), pp. 33–59.

49. According to John Gillingham, 'In essence everything that sixteenth-century Englishmen believed about Ireland' observes John Gillingham, 'can be found in the writings of Gerald of Wales'. See Gillingham, 'Images of Ireland 1170–1600: The Origins of English Colonialism', *History Today*, 37 (1987), p. 18. In a famous passage in Gerald's work, the author writes: 'They are a wild and inhospitable people. They live on beasts only and live like beasts. They have not progressed at all from the primitive habits of pastoral living'. Gerald quoted in *Great Deeds in Ireland: Richard Stanihurst's De Rebus in Hibernia Gestis*, eds. John Barry and Hiram Morgan (Cork: Cork University Press, 2013), p. 467, n. 94.

50. Giraldus Cambrensis, *The Topography of Ireland*, trans. Thomas Forester and revised and edited with additional notes by Thomas Wright (Cambridge: Medieval Latin Series, 2000), p. 4.

51. Cambrensis, *The Topography of Ireland*, pp. 4–5.

52. Thomas O'Loughlin, 'An Early Thirteenth-Century Map in Dublin: A Window into the World of Giraldus Cambrensis', *Imago Mundi*, 51 (1999), p. 32.

53. Spenser, *A View of the State of Ireland*, p. 11.

54. Warham St. Leger quoted in J.H. Andrews, *Plantation Acres: An Historical Study of the Irish Land Surveyor and His Maps* (Belfast: Ulster Historical Foundation, 1985), p. 28.

55. Bernhard Klein, 'English Cartographers and the Mapping of Ireland in the Early Modern Period', *Journal for the Study of British Cultures*, 2:2 (1995), p. 131. See also J.H. Andrews, *The Queen's Last Map-Maker: Richard Bartlett in Ireland, 1600–3* (Dublin: Geography Publications, 2008).

56. John Davies quoted in Mark Netzloff, 'Forgetting the Ulster Plantation: John Speed's *The Theatre of the Empire of Great Britain* (1611) and the Colonial Archive', *Journal of Medieval and Early Modern Studies*, 31:2 (2001), p. 321.

57. For a more extensive examination of Norden and Speed's Irish maps, see Andrew Bonar Law, *The Printed Maps of Ireland to 1612* (New Jersey: Eagle Press, 1983), and J.H. Andrews, 'John Norden's Maps of Ireland', *Proceedings of the Royal Irish Academy*, 100C:5 (2000), pp. 159–206.

58. Henry Sidney quoted in Baker, 'Off the Map: Charting Uncertainty in Renaissance Ireland', p. 83.

59. Retha M. Warnicke, 'Nowell, Laurence (1530–c.1570)', *ODNB* (Oxford: Oxford University Press, 2004), http://www.oxforddnb.com/view/article/69731 (accessed 27th June 2020).

60. Andrews, *Plantation Acres*, pp. 28–29.

61. Breen, 'Spenser's "Imaginatiue Groundplot"', p. 151.

62. See, for example, J.H. Andrews, *Shapes of Ireland: Maps and Their Makers 1564–1839* (Dublin: Geography Publications, 1997) and Netzloff, 'Forgetting the Ulster Plantation', pp. 313–348.
63. William J. Smyth, *Map-making, Landscapes and Memory* (Notre Dame, Illinois: University of Notre Dame Press, 2006), p. 21.
64. Annaleigh Margey, 'Visualising the Plantation: Mapping the Changing Face of Ulster', *History Ireland*, 17:6 (2009), p. 43.
65. Annaleigh Margey and J.H. Andrews, 'A Hitherto Unknown Sketch Map by Lord Burghley', *Imago Mundi*, 64.1 (2012), p. 96.
66. John Breen, 'Spenser's "Imaginatiue Groundplot": A *View of the Present State of Ireland*', *Spenser Studies*, 12 (1991), p. 165.
67. Avery, 'Mapping the Irish Other', p. 269.
68. Avery, 'Mapping the Irish Other', p. 263.
69. Avery, 'Mapping the Irish Other', p. 263.
70. Joanne Woolway Grenfell, 'Significant Spaces in Edmund Spenser's *View of the Present State of Ireland*', *EMLS*, 4:2 (1998), para. 7.
71. Grenfell, 'Significant Spaces', para. 8.
72. Grenfell, 'Significant Spaces', para. 17.
73. Avery, 'Mapping the Irish Other', p. 268.
74. Bruce McLeod, *The Geography of Empire in English Literature: 1580–1745* (Cambridge: Cambridge University Press, 1999), p. 32. For more on Spenser's manifesto for the English in Ireland, see Joan Fitzpatrick, *Shakespeare, Spenser and the Contours of Britain: Reshaping the Atlantic Archipelago* (Hertfordshire: University of Hertfordshire Press, 2004), pp. 26–57.
75. John Norden, *John Norden's The Surveyor's Dialogue (1618)* ed. Mark Netzloff, (Surrey: Ashgate Publishing, Ltd., 2010), p. 21.
76. Norden, *The Surveyor's Dialogue*, p. xv.
77. Another famous example of the English tendency to imagine Ireland imagistically is found in John Speed's *Theatre of the* Empire *of Great Britaine* (1611). In the section of the atlas dedicated to 'The Kingdome of Irland', Speed depicts a land fully under the control of English forces, cartographic domination reciprocating militaristic occupation. 'Speed', J.H. Andrews writes, 'shows a province under subjugation, its forests (what is left of them) divided by passes, a warship patrolling its largest lake [Lough Neagh], its rebellious spirit curbed by a network of military strongholds' (Andrews, *Shapes of Ireland*, p. 107). This depiction is maintained in other aspects of the representation. As well as a highly detailed map of the country, Speed also includes pictorial depictions of the various strata of Irish society, from 'New English' to 'Old English' (descendants of the first invasion of Ireland in the twelfth century) and also 'Native Irish', complete with mantle recalling Irenius's description in *A View*. See Jerry Brotton, 'Mapping the Early Modern Nation: Cartography Along the English Margins', *Paragraph*, 19:2 (1996), pp. 139–155.
78. Lupton, 'Mapping Mutability: Or, Spenser's Irish plot', p. 95.
79. David Woodward, 'The Image of the Map in the Renaissance', *Approaches and Challenges in a Worldwide History of Cartography*, eds. David Woodward, et al. (Barcelona: Institut Cartogràfic de Catalunya, 2000), p. 136.
80. Woodward, 'The Image of the Map in the Renaissance', p. 136.
81. Spenser, *A View of the State of Ireland*, p. 96. Willy Maley has reflected on the double implications in this passage, noting what he describes as 'probably the most classic example of Spenser's propensity for wordplay'. In particular, Maley addresses Eudoxus's intention to 'lay [the map] before me, and make mine eyes (in the meane time) my Schoolemasters, to guide my understanding of your plot', observing that the sequence under discussion 'leaves one

wondering whether the *View*, without woodcuts or typographic design, is yet the most visual of Spenser's text'. See Maley, *Salvaging Spenser: Colonialism, Culture and Identity* (London: Macmillan Press, 1997), pp. 79–80.

82. Spenser, *A View of the State of Ireland*, p. 29.
83. Nandini Das, *Renaissance Romance: The Transformation of English Prose Fiction, 1570–1620* (Surrey: Ashgate, 2011), pp. 73–74.
84. *A Concordance to the Poems of Edmund Spenser*, ed. C.G. Osgood (Washington: Carnegie Institution of Washington, 1969), p. 655.
85. Spenser, *A View of the State of Ireland*, pp. 22, 91, 100.
86. Spenser, *A View of the State of Ireland*, p. 95.
87. McLeod, *The Geography of Empire in English Literature*, p. 33.
88. Maley, *Salvaging Spenser*, p. 80.
89. Palmer, *Language and Conquest*, p. 40.

Part II

'What is cosmographie?'

Teaching the science of
world describing

4 'There is none so good lernynge'

Cartography and cartographic instruments in early modern English educational treatises

Having considered the early modern map and its various employments in the study, frontispiece and colony, this chapter examines cartography in the context of the teaching of geographical science in sixteenth- and seventeenth-century England. It examines how educational theorists engaged with the map and conceptualised its function as an educational aid. In doing so, it will provide a wider context for those mapping moments present in Spenser, Speed and Drayton, which utilise cartography's varied representative capacities. It explores some of the definitional complexities surrounding the use of the term 'geography' in Renaissance humanist education. It also explores how some of the ideas that emerged from the cartographic enthusiasm of early modern intellectuals, alongside more profound notions of humanist enquiry, influenced the way maps, globes and mapping instruments were used in the learning environment, bringing cartography's multivalent physical and symbolic characteristics to the fore.

'What is geographie?': A problematic definition

In a letter written in January 1518 to Anton Van Bergen, the youngest son of the councillor to Margaret of Austria, courtier to Henry VIII and eventual governor of Luxembourg Jan Van Bergen, the influential intellectual Desiderius Erasmus advises that 'History will mean more to you if you have a taste of geography first'.[1] Erasmus's advice was reiterated by other humanist scholars: 'Geographical knowledge', Jonathan M. Smith states, 'was one of the elements added to a classical education by humanist writers such as [...] Leonardo Bruni, Leon Battista Alberti, and Juan Luis Vives'.[2] A primary factor motivating this call for the integration of geographic science into the scholarly syllabus was the appearance of the printing press. As with many other aspects of early modern epistemology, the advancement of print technology had a profound effect on the status of the world describing.[3] Abundance generated attention: the flourishing of the printing press and its resultant profusion of geographic texts engendered a greater emphasis on the place of geography within the humanist scholarly curriculum. J.N.L.

DOI: 10.4324/9781003200376-7

Baker, in his survey of geographic textbooks used in English education, writes:

> With the expansion of knowledge in the fifteenth and sixteenth centuries one of the disadvantages from which geography had hitherto suffered was removed though the speed of the advance brought with it many difficulties. Simultaneously the printing of the works of classical geographers, such as Strabo, Ptolemy, and Mela, stimulated geographical thought. It is not, therefore, surprising to find the educational experts of the new age insisting on the value of geographical teaching.[4]

The arrival of the Gutenberg press in 1432 reinvigorated geographic teaching through the facilitation of the mass reproduction of recuperated ancient texts such as Strabo's *Geographica* and Ptolemy's *Geography*. The reproduction of these works had a dramatic influence upon many geographical practices: the re-emergence of the *Geography,* for instance, led to the appropriation of Ptolemaic spherical projection by European cartographers and its establishment as the normative descriptive mode in cartography from the sixteenth century onwards.[5]

However, despite the expanding geographic literature of the period, a prevailing categorical definition of geographical science is not easy to find. What constituted the pursuit, rubric and aims of geographical, and more broadly cosmographical enquiry (study of the cosmos) was much debated. As S.K. Heninger observes, citing the interrogatory sequence in Thomas Blundeville's *His Exercises* (1594), where questions such as 'What is Cosmography?', 'What is Astrologie?' and 'What is Geographie?' are posed and answered:

> Because of its all-inclusiveness, cosmography was necessarily divided into several specialized sciences, defined according to the direction of emphasis, toward heaven or toward the earth. There was astronomy, which turned its gaze on the celestial bodies; geography, which concentrated on the terrestrial globe; and astrology, which concerned itself with the operative relations between the two. A fourth science, chorography, was often further defined because, in contrast to geography, it described the part divorced from the whole and depended upon particular measurement of an area without reference to the globe. Its intention was locally descriptive rather than mathematical and universal.[6]

Heninger's summary provides a sensitive detailing of one description of the intricate structures of early modern English geographical science. Incorporated into the study of the cosmos (cosmography) were four separate sub-disciplines. Astronomy and geography focused on the heavens and the earth, respectively, seeking to describe and map out their terrains. Astrology, meanwhile, postulated a link between the heavens and the earth and sought to discern a greater understanding of such connections. Lastly, chorography retained a focus on the local, considering and detailing individual delimited topographical spaces.

Yet while some writers such as Blundeville employed these definitions, such disciplinary demarcations were by no means universal. At the boundaries of the sub-disciplines, there was slippage in definition, process and purpose. As Heninger acknowledges, 'In practice, writings in any one of these four fields impinged upon all of them'.[7] Likewise, Lesley Cormack has noted that even well into the seventeenth century, 'there continued to be a large amount of overlap' between the tripartite 'categories of mathematical, descriptive, and chorographical geography'.[8] In the endeavours of cosmography, geography, astrology, astronomy and chorography subject, structure and definition were all subject to interpretation and variance.

Drawing out these interpretations and variances is essential to understanding how the subject was regarded and consequently taught in the period. James Dougal Fleming, in an addendum to a collection of essays based around the theme of the 'invention of discovery' has suggested that for scholars examining the early modern period and its discourses and disciplines 'the issue [...] is not just how to understand a given historical object [...]; but to understand'.[9] Outwardly concerned with analysing the 'early-modern emergence of natural science', Fleming's argument nonetheless has relevance for our attempt to grasp what geography meant to intellectuals of the time:

> For the object in question – the early-modern emergence of natural science – itself entails certain fundamental yet questionable claims about what counts as understanding. Therefore, studying the early-modern period, with the emergence of modern natural science at its core, means doing hermeneutics: trying to understand understanding.[10]

'[D]oing hermeneutics' in relation to early modern geography throws up many challenges. A primary difficulty is the fact that the very notion of a specific 'scientific "disciplinary" geography' would have been an extrinsic idea to sixteenth- and seventeenth-century scholarship, both on an individual and also on an institutional level. While the subject of geography was inaugurated at England's two most substantial educational institutions (Oxford and Cambridge) in the 1540s, this should not be taken as a sign of standardisation. 'Geography was being taught in academic institutions in the 1580s', write Robert Mayhew and Charles Withers, 'yet was not departmentalized until the 1880s'.[11] Due to this lack of 'departmentalization', the subject lacked many of the easily identifiable features of a modern institutionalised discipline, such as formalised investigative approach, defined subject area and structured hierarchies of qualification.[12]

Conscious of such lacunae, Withers and Mayhew propose that 'we should expect that geography was taught [in early modern universities] to the extent that it was germane to a degree in the arts'.[13] Both call attention to the connectedness between the subject and degree classification (geography and the arts) in universities and suggest it emerged out of the humanist educational ideal of a multi-disciplinary learning.[14] As the case of the playwright Christopher Marlowe – who studied geography as part of his Master of Arts degree at Corpus Christi Cambridge – illustrates, even classifying geography as a specified branch of early modern *science* is problematic. An effect of the holistic educational mode recommended by humanist thinkers such as Erasmus, Michel de Montaigne and Francis Bacon is that our own twenty-first-century disciplinary definitions, with their emphasis on distinction and categorisation, are anachronistic.[15] The propagation of eclecticism in educational theory meant that, as with anatomy, investigative approaches learned in one discipline were actively encouraged in another. William Pemble's *A Briefe Introduction to Geography* (1630) forcefully demonstrates the difficulty of even situating the subject within the current prevailing scholarly binaries of 'humanities and sciences'. 'Geography', writes the author in the opening chapter of his textbook, 'is an art or science teaching us the generall description of the whole earth'.[16] Master of Arts at Magdalen Hall, Oxford, Pemble's definition provides an insight into the classificatory fluidity of the topic in academia in the period.

Geographical knowledge and its limits

With such definitional ambiguities in mind, how then should we approach the topic of early modern 'geographical knowledge' with a view to interrogating its mapping manifestations? A starting point would be to recognise its multiplicity, both in focus and approach. In attempting to delineate geography, or rather what contemporaneous intellectuals understood as geography (in Fleming's succinct phrase attempting to 'understand understanding'), we must first acknowledge that the endeavour was, alongside natural philosophy, one of the most capacious scholarly pursuits in Renaissance Europe.[17] It encompassed chorography, cosmology and astronomy, and took in practical (surveying) as well as the theoretical (geometry) pursuits. Its mediums were often cartographic but could also be textual and diagrammatic. In many respects, this diversity is maintained in the twenty-first-century university, geographical courses at university reflecting the etymology of the term (*geo-* 'earth', *graphia-* 'describe-write') to encompass an array of diverse topics and methods.[18] However, even in comparison to the panoramic scope of modern geography, early modern 'geographical knowledge' retains a striking diversity in focus, variety of classificatory designations, fluidity of structures and multiplicity of theoretical practices.

This was particularly true in an educational context, as many writers provided conflicting definitions of the very fundamentals of the discipline. '[G]eography as taught in British universities before 1887', write Withers and Mayhew, 'had important connections with other subjects and different meanings and different geographies according to the individuals and institutional settings involved'.[19] For the twenty-first century geographical student, the subject (despite its many branches and sub-disciplines) is summarised by its *OED* entry:

> [Geography is t]he field of study concerned with the physical features of the earth and its atmosphere, and with human activity as it affects and is affected by these, including the distribution of populations and resources and political and economic activities.

Strikingly, to some readers, even the most basic part of the subject – what geography was about – was open to interpretation. While today, as the *OED* demonstrates, the study of geography is accepted as being 'concerned with the physical features of the earth and its atmosphere' and 'human activity', in the period under examination, such obvious connections were not so distinct. The apparently indelible link between geography and the earth was often open to interpretation. Many thinkers agreed that the focus of geography and geographers should be terrestrial, with the declaration in Luis de Granada's *A memoriall of a Christian life* (translated in 1586) that the subject is 'a description of the whole Earth, as much as is discovered to us' being a typical classification. In *Geographie delineated forth in two bookes* (1635), Nathaniel Carpenter, a teacher of geography at Exeter College, Oxford, also stresses the subject's etymological roots – 'The Nature of *Geographie* is well expressed in the name: For *Geographie* resolued according to the *Greeke* Etymologie, signifieth as much as a description of the Earth'.[20] Such instances corroborate to an extent the claim that geography 'was restricted to the scale of the earth as a whole'.[21] However, countering this assertion by Robert Mayhew is the rare, though nonetheless notable association of geography with the study of the moon: in his *Anatomy of Melancholy* (1621), Robert Burton describes the work of the eminent German astronomer Johannes Kepler as 'Lunar Geography'.[22]

This divergent application of the disciplinary categorisation carries particular resonance because of his background. Burton had an active engagement with the teaching of geography at the university level in the late 1590s – according to J.N.L. Baker, it is 'possible to include' Burton as '[a]mong early teachers of geography in Oxford'.[23] The author maintained a mindfulness of contemporary 'geographical knowledge' throughout his life and was in possession of 'a good library of geographical books [including] Mercator's *Atlas Geografer* and Ortelius' *Theatrum Mundi* as well as a number of surveying books and instruments'.[24] Furthermore, Burton's *alma mater* at Oxford, Christ Church, was perhaps the most vibrant arena of geographical learning in late-sixteenth-century England, attracting and nurturing several important figures in the fields of navigation, history and mathematics including Richard Hakluyt (author of the renowned *Principle Navigations* (1589)) and William Camden, the antiquarian, chorographer and topographer.[25] Burton's yoking of astronomical observance to geography represents something of an exception among definitions of the subject; but it also shows that the attention of geographical inquiry, even among those sensitive to the evolving discourse of 'geographie', was not always fixed on the earth.

Another important fact to recognise in attempting to formulate an understanding of early modern geography is the indeterminacy regarding exactly how 'geographical knowledge' was structured. Many, like Thomas Blundeville, recognised geography as a sub-stratum of the wider discipline of 'cosmographie'. The aim of such a subject is perhaps best expressed in John Dee's important 1570 translation of the Greek mathematician Euclid:

> Cosmographie, is the whole and perfect description of the heauenly, and also elementall parte of the world, and their homologall application, and mutuall collation necessarie. This Art, requireth Astronomie, Geographie, Hydrographie and Musike. Therfore, it is no small Arte, nor so simple, as in common practise, it is (slightly) considered.[26]

In Dee's model, geography formed part of a wider formation of subjects alongside astronomy, hydrography and even music specifically designed to describe the heavens and earth.

And yet these sub-topics in themselves retained a definitional ambiguity, often overlapping in focus. The lack of disciplinary articulation can be best demonstrated by examining the practice of chorography: 'Geography writ small', chorography was 'fundamentally practical since it emphasized the surveying of estates close to home, the compilation of genealogies of local families, and the description of local attractions and commodities'.[27] As Lucia Nuti shows, endeavouring to describe regions or localised areas was an important part of the discourse and accrued its significance from its specific rather than general, and qualitative rather than quantitative emphasis:

> Chorography was [...] intended to convey only a limited knowledge of the earth: an individual part rather than the whole, the secondary rather than the primary, quality rather than quantity.[28]

In geographical nomenclature, chorography represented a part of the discipline of geography as much as a county or shire represented part of the country.[29] Such an analogy is evidenced by reference to prominent literary geographies: in Speed's *The Theatre of the Empire of Greate Britain* (1611), the author presents to the reader 'an exact geography of the kingdomes of England, Scotland, Ireland, and the iles adioyning'.[30] Constituting this 'geography' is an admixture of cartography and chorography. Each county subsection includes a map displaying the land alongside a textual description outlining its features, history and other sundry information. In Speed's conceptualisation, chorography represents a part of the discipline of 'geography', in the process concurring with the definition given by John Dee: 'Chorographie seemeth to be an underling, and a twig, of Geographie'.[31] Chorography, in Dee and Speed's understanding, is explicitly incorporated into the pursuit of geography, much as geography forms a sub-discipline of cosmography. However, hydrography or the description of bodies of water frequently featured in texts which fashioned themselves as chorographical: Michael Drayton's *Poly-Olbion* (1612) includes descriptions of the Thames (Song XVIII, 97–108). Thus, the distinction Dee makes between geography (and its subdiscipline chorography) and hydrography was not always consistently applied as writers consciously played with the classificatory boundaries of world-describing.

Further divergent classifications pervaded. Many writers in the period often drew a distinction between geography and chorography as separate subjects and even classified the two disciplines as equal in the overall epistemological arrangement of cosmography. One notable example of this is Henry Peacham's *The Compleat Gentleman* (1622), a popular educational treatise. Peacham, contrary to Dee and Speed, defines chorography not as a part of geography but rather as a separate, though related discipline, writing that 'Cosmography containeth Astronomie, Astrologie, Geography and Chorography'.[32]

Defining the map and the mapper

Chorography's indeterminate status emblematises a fluidity of definition and complexity in the practice of geographical science in the late sixteenth and early seventeenth centuries. Similarly, the status of cartography was often nebulous, existing both as a singular pursuit and also a supplementary part of a larger subject. The trend for a fluid and expansive approach to practice can be detected most clearly in the sphere of publishing. Wholly cartographic texts were rare in the period – many works purporting to be 'geographical', like Speed's *Theatre*, presented to the reader an amalgamation of text and image with chorographical or topographical description accompanying map illustration. The most widely read atlases in the period were mainly of continental origin, namely Abraham Ortelius's *Theatrum Orbis Terrarum* (1570) and Gerard Mercator's *Atlas* (1578). At once signifying the pre-eminence of northern European continental cartography in the production of maps in the sixteenth and seventeenth centuries, this also indicates the wide disciplinarity of English geographical publications and its susceptibility to other seemingly discrete disciplines such as antiquarianism, history and geology. While Christopher Saxton's collection of English and Welsh county maps, published in 1579, proved popular, gained for the author 'considerable patronage and assistance, and also significant reward, including Queen Elizabeth I's favour'.[33] The novelty of Saxton's predominantly cartographic survey marks out its uniqueness in an area proliferated by multimedia and multidisciplinary texts.

Where maps were included in texts, authors and publishers frequently used cartographic representations as accompaniments to chorographical descriptions. Cartography was not conceived of as a singular discipline worthy of a whole publication but rather part of a wider subject of interest in geography. For example, when William Camden published *Britannia* in 1586, the author entitled his work 'Britain, or A chorographical description of the most flourishing kingdoms, England, Scotland, and Ireland, and the islands adjoining, out of the depth of antiquity beautified with mappes of the several shires of England'.[34] Camden characterises maps as decorative appurtenances to the main element of the book, namely chorographical description in the form of text: the maps are intended to 'beautify' the chorography, providing an aesthetic supplement to the textual element of his book.

A further indicator of the indeterminacy of the discipline is its use as a generic classification in geographic texts. The title of William Cuningham's *The Cosmographical Glasse* (1559) is illustrative in this regard. Promising the reader a guide to the 'pleasant principles of cosmographie, geographie, hydrographie, or nauigation', Cuningham implicitly posits an equal status between 'geographie' and 'cosmographie'.[35] This contrasts with the classification that 'cosmographie consisteth of geographie' as expressed by Dee and others, underlining the fact that geography, cosmography and other similar sciences fluctuated in their perceived parameters and status.

Using maps and globes to teach geography

Geographical knowledge then was a capacious subject, with a widely defined, often ambiguous rubric and fluid epistemological focus. Nonetheless, in spite of such varying interpretative structures, geography developed as a prominent component of the early modern humanist educational curriculum. This burgeoning interest was particularly marked among university students: according to Lesley Cormack in her survey of geographic teaching from 1500 onwards, 'From the grammar schools on, both formal and informal educational systems had some interest in the study of the cosmos'.[36] As the wealth of geographic texts published from the 1530s onwards and the frequent re-issuing of many of the most prominent volumes in this genre demonstrates, geography was a popular academic endeavour. Moreover, its persistence in popularity in spite of an uncertain and shifting definitional landscape demonstrates the perceived possibilities afforded by geography in the cultural imagination as a scholarly and leisurely pursuit and also its multiple applications in a broad range of scholarly spheres.

One of the most prominent aspects of the developing interest in geography was the advocacy of an engagement and familiarity with the physical accoutrements of the discipline such as maps, compasses and globes. Among contemporary educational treatises such as Thomas Elyot's *The Boke Named the Governor* (1531), Thomas Blundeville's *Exercises* (1597) and

Henry Peacham's *The Compleat Gentleman* (1622), there is a continual emphasis on the need for the presence of cartographic equipment in the classroom and study. This reflected the inherent practicality of the discipline. By many pedagogists, cartographic objects were seen not only as things to delight and imaginatively transport – but they were also understood as vital devices for learning about geography and its related disciplines. Furthermore, the theory/practice nexus at the heart of geographical science was emblematic of the more profound humanist educational paradigm that stressed both theoretical book learning and functional experimentation.

The proliferation of cartographic documents and instruments in the early modern classroom occurred within a broader educational circumstance that promoted empirical modes of learning. Cormack notes that while 'humanist pedagogy aimed to teach Latin texts of antiquity through rote', in other subjects, 'humanists stressed the need to teach with a light hand and to use games and playfulness rather than the rod as an enticement to learning'.[37] Likewise, Elly Dekker emphasises the humanist stress on learning by demonstration and applied engagement. Arguing that 'the uses of demonstration models in university education is a phenomenon that cannot be viewed separately from the trend towards hiring specialists for teaching mathematics, astronomy, and geography at universities', Dekker has detailed some of the earliest uses of cartographic instruments such as the armillary sphere, which represented the lines of celestial longitude and latitude and other astronomically important features, and the celestial globe, which marked out the night sky on a spherical survey, in European education.[38] Archetypal is the influential work of the German scholar and teacher Peter Apian: '[Apian's] complete oeuvre', writes Dekker, 'is interspersed with all sorts of demonstration models of wood and paper in two and three dimensions to overcome mathematical barriers'. Referring to a particular illustration entitled 'De speculo cosmograph (Cosmographic glass)' included in Apian's *Cosmographicus liber* (1524), Dekker describes how the author utilised cartographic objects for instruction:

> With the help of Apian's paper instrument, a variety of cosmographic problems could be solved: locating a place on earth once the geographical longitude and latitude of a place was known; familiarizing students with the use of spherical co-ordinates; working out the relation between local times at different places, in which case the instrument served as an analogue computer; or finding where on the earth the sun appears at the zenith on certain days of the year to explain the concept of the zones.[39]

Typifying the advocacy characteristic of many geographers, Apian's instrument highlights the benefits arising from a physical engagement with cartographic objects.[40] Maps and globes enabled pupils to carry out problem-solving exercises, such as applying latitude to locate a particular place or locating the position of the sun in relation to the earth at a particular time of day. By these means, they would familiarise themselves with and comprehend the validity and accuracy of certain mathematical, geometrical and geographical principles. Becoming accustomed to features such as co-ordinates, time zones and solar trajectory served a similar purpose. Geographical knowledge would be imparted not only through dictation but also through practical exercise.

English writers reciprocated Apian's stress on the practical implementation of learned principles as geography in the classroom and, in the latter half of the sixteenth century, sought to foreground a mutually beneficial exchange between theory and practice. In teaching geography within a model where knowledge was attained from the application of theoretical principles, the subject inculcated into students a practicality vital to many different and divergent professions. Noting that '[g]eographers had to develop a theoretical superstructure for their knowledge', Lesley Cormack sets out the benefits of geography's dual symbiosis of learning for artisans and scholars alike:

> At the same time, [geographers'] knowledge could never be complete without reliance on the skill and information of navigators and travelers, or of personal experience. [...] [G]eographers, who were interested in increasing and developing their knowledge of the world, had to be theoretical, practical and political. Men interested in geography developed multiple and overlapping roles as scholars, craftsmen, and statesmen, or various combinations of the three. The investigation of geography in the late sixteenth century embodied that dynamic tension between the world of the scholar, since geography was clearly an academic subject legitimated by its classical, theoretical, and mathematical roots, and the world of the artisan, since it was inexorably linked with economic, nationalistic and practical endeavors.[41]

An illustration of this 'fruitful exchange between the life of the mind and that of the marketplace' in the discipline of geography is the collaboration of the 'university-trained mathematician' John Dee with the merchant and translator Henry Billingsley in his translation of Euclid.[42] In his introduction, Billingsley, with an eye to the possibilities afforded by 'a perfect knowledge of Geometrie' writes, 'Many other artes also there are which beautifie the minde of man: but of all other none do more garnishe & beautifie it, then those artes which are called Mathematicall'.[43] Both Dee and Billingsley were familiar with the benefits of geometry and mathematics not just to geography but also in other disciplines. In exhibiting an awareness of those readers who approached their book from various disciplines and more craft-orientated professions beyond theoretical mathematics, the translation of Euclid demonstrates the responsiveness of writers to the varied character and potentially multifunctional role of geographical learning.

In a similar vein is *Pantometria*, published a year later by the mathematician Leonard Digges. A three-volume treatise on mathematical geography or 'Geographical Practise [...] containing Rules manifolde for mensuration of all lines, Superficies and Solides', Digges's study was completed by his son Thomas and prefaced with an address to those 'either of ignoraunce or malice [that] shoulde affirme it [geometry] unprofitable'. Referring to Plato, the younger Digges emphasises the many branches of knowledge open to those with an understanding of the mathematical principles that underpin geometry and, expressly, geography:

> In sundrie other his works also of naturall Philosophie, as the Physikes, Meteores, de Coelo & Mundo. &c. ye shal finde sundrye Demonstrations, that without Geometrie may not possibly be understanded. And to leue Philosophie, how necessarie it is to attayne exacte knowledge in Astronomie, Musike, Perspectiue, Cosmographie and Nauigation, with many other Sciences and faculties, who so meanely trauayleth therein shall soone finde.[44]

Digges's introduction conveys the intertwining schematics of early modern knowledge systems. Chiefly, practical disciplines such as navigation are described as having their foundation in geometrical principles. However, Digges expands this multidisciplinary connectivity: in an echo of the connotative categorisation of geography as an 'art or science', superficially 'creative' subjects such as music and more oblique subjects such as perspective are also intimately linked with the same theoretical structures of knowledge that undergird geography and its own sundry sub-disciplines. While acknowledging that Digges' stress on the interdependency of knowledge pursuits may be in part motivated by personal interest – he was, after all, writing in a book published with the intention of teaching geometry, and hence his stress on the importance of the subject would be understandable – the writer nonetheless maps out a multifaceted epistemological landscape. In such a fluid epistemological context, the language, nomenclature and structure of 'geographical knowledge' involved an exchange not only between theory and practice but also between the practical application of geography and its usefulness for 'many other Sciences and faculties'.

In seeking to define geography and its utility, writers tacitly made an exemplar of geography in the advancement of the idealised multidisciplinary humanist education. As Digges' preface suggests, the advancement of the polymath was in part inspired by a belief that different types of knowledge thrived in relation to each other. William Poole has provided an illuminating outline of the fluidity of contemporary intellectualism in this regard. Citing the work of John Aubrey, Poole identifies the Shakespearean biographer as stereotypical of the 'eclectic' learning of the scholar: '[Aubrey] was really an interested eclectic, a man who was an amateur in most of his interests, profound in a few of them, and always exercised by how one kind of discussion might lead on to another'.[45] This sense of 'one kind of discussion leading onto another' permeates Digges's introduction and geographic discourse more generally from Dee to Saxton to Speed. Natural philosophy and 'Physikes' necessitate geometry, while in turn, 'Astronomie, Musike, Perspectiue, Cosmographie and Nauigation' are predicates of geometry. The scholar who 'so meanely trauayleth therein shall soone finde'. Fashioned as a traveller, Digges' reader can explore the topography of knowledge, roving across the superficial disciplinary boundaries to gain a greater comprehension of the whole lay of the land.

Why use maps?

Geography then emerged as a key prototype of early modern polymathy. While representative of the eclectic inclinations of early modern educational theorists, pedagogic texts promoting geographical learning frequently emphasised the fecundity of geographical science as a practical and theoretical activity. As university numbers grew in England in the mid-to-late 1500s, the predominance of theological degrees slowly dissipated and post-school education was pursued by many as advantageous to finding a career in diverse spheres.[46] Jonathan M. Smith has outlined the attraction of university for many young men seeking employment in positions of influence and – more meaningfully for our study of the teaching of geography – the centrality of geographical science to an increasingly practically inclined, vocationally tailored curriculum. Quoting Thomas Hobbes's classification of geography in *Leviathian* (1651), Smith points out that:

> 'knowledge of the face of the earth' assumed its modern importance because the size and geographical complexity of states increased and that attainment of this knowledge was, as it remains, a prerequisite for access to political power.[47]

'A prerequisite for access to political power', according to Smith, geography and its ancillary discipline of cartography were viewed as a critical factor in the enterprise and success of the aspiring gentleman.

As a consequence, maps and globes entered the school and university curriculum, proliferating widely in scholarly environments, and were utilised in different ways. As Cormack observes:

> [T]hey performed different functions and had a different status in each setting. [...] [M]aps seen as useful only for their illustrative properties, informal market-driven and patronage-supported education privileged map knowledge as necessary for the gentleman or merchant. Increasingly during this period, those in positions of power saw maps as important imperial, mercantile, and aesthetic objects, and this

encouraged an informal education system, which relied on their patronage, to introduce maps into education.[48]

Exploration of the advancement of cartographic objects in the education of young scholars also provides us with an understanding of how map readers and cartographic enthusiasts engaged with the discipline of cartography and its sundry instrumental, theoretical and conceptual accoutrements. Maps and globes, many educational theorists realised, had tangible benefits in the education of young scholars and were frequently assimilated into the pursuit of certain educational aims. By interrogating these assimilations and the strategies underlying them, we can determine some of the perceived applications of cartography in the period.

Such an approach must be tempered by the acknowledgement that any historical comprehension of the role of cartography in the early modern classroom is, somewhat paradoxically, based largely on theoretical texts. If the multiple editions of those treatises most unequivocal in their backing of a physical familiarity with cartography are descriptive, scholars were, in theory, accustomed to the practical discipline of mapmaking.[49] However, evidence for the realities of the usage of cartographic documents and instruments in the teaching environment of the period is sparse.[50] Aspects such as the student's familiarity with maps and globes, how they were used in lessons or even where they were situated in the spatial configuration of the learning space are largely absent from the historical record. Some scholars have even used such evidential absence as part of a challenge to the established historical orthodoxy of a cartographic boom in English culture, questioning both the extent of the proliferation of maps and the degree of cartographic literacy among the general population in the period.[51]

Yet, in those theoretical tracts, vital indicators or practices are evident. Educational treatises and school textbooks of the period which support a practical engagement with the objects of cartography and mapping, some of the fundamental ideas which underpinned the ways in which intellectuals engaged with the map are outlined. Of particular note is the reciprocation of the cartographic enthusiast's fascination with the physical properties of the map – its colour, variety and seeming ability to reduce the world to a single vista. In part, this may be attributed to the fact that many of the most substantial educational theorists were themselves passionate map readers. One of the most effusive celebrants of the joy of maps, Thomas Elyot, was also one of the most ardent proponents of the use of cartographic objects in the classroom. In *The Boke Named the Governour* (1531), Elyot writes:

Also to prepare the chylde to vndestandynge of histories, whiche beinge replenished with the names of countres & townes vnknowen to the reder, do make the historye tedious, orels the lesse pleasaunt, so if they be in any wyse knowen, it encreaseth an inexplicable delectation: It

shalbe therfore, and also for refresshynge the wytte, a conuenyent lesson, to beholde the olde tables of Ptholomee, wherin all the worlde is
painted, hauinge fyrste some introduction in to the sphere, wherof nowe
of late be made very good treatises, and more plaine and easy to lerne
than was wonte to be. All be it there is none soo good lernynge, as the
demontration of cosmographie, by materyall fygures and instrumentes,
hauynge a good instructour. And surely this lesson is bothe pleasaunt
and necessary.[52]

Elyot's promotion of the 'demontration of cosmographie, by materyall
fygures and instrumentes' bases itself upon several premises, the first of
which is the recognition of the growing interest and contemporaneous burgeoning sophistication and increasing dissemination of cosmographical,
geographical and cartographical textbooks.[53] *The Boke*, in its celebratory
rhetoric, actively seeks to engage with this burgeoning discourse. The author
recommends that the student have 'some introduction in to the sphere [or
globe], wherof nowe of late be made very good treatises, and more plaine
and easy to lerne than was wonte to be'.[54] Moreover, Elyot takes care to
affirm the efficacy of practical engagement with the objects of cartography:
earlier in his treatise Elyot declares, 'In whiche studies I dare affyrme, a
man shal more profyte in one wike by figures and cartis, well and perfect-
ely made, than he shall by the onely redyng or herynge the rules of that
scyence, by the space of halfe a yere at the leaste'.[55] Distinguishing theory
from practice, Elyot rails against didactic dictation or 'the onely redyng or
herynge the rules of that scyence'. 'Figures and cartis [charts], well and perfectly made', serve to stimulate the mind and accelerate the learning process
exponentially.

The Boke's endorsement of a practical engagement with cartographic
objects again underlines the perceived enjoyment of map reading: in a reference which recalls the sheer delight of map reading to the early modern
cartographic consciousness, Elyot refers to the personal gratification that
can be derived from studying cartographic objects. One of the most immediate benefits of the use of cartographic objects in education is the visual
variety they introduce in an educational setting. Emphasised is 'demontration [...] by materyall fygures and instrumentes' as pleasurable learning.
The student for whom studies seem 'tedious, orels the less pleaseaunt'
may find their subject 'more plaine and easy to lerne than was wonte to
be'. 'Refresshynge the wytte', maps and globes, tables and charts break up
the monotony of learning by rote geographical facts such as 'the names of
countres & townes'. In short, if such 'materyall fygures and instrumentes'
are incorporated into a lesson, that lesson becomes not only 'necessary', but
also 'pleasaunt'.

As set out in the introductory chapter, many in the main embraced the
delights induced by the contemplation of the map, eulogising upon its col-
ourfulness, variety and seemingly endless diversity. The propinquity of

Elyot's encomium on the delights of map study and his recommendation of practical 'demontration' reveals the concomitance of cartographic enthusiasm to the usage of maps and globes in education. *The Boke* exhibits one important instance of the conjuncture of developing educational theories surrounding the teaching of geographical science and cosmography and the period's fascination with maps and globes as physical objects.

The aforementioned allusion to Plato in *Pantometria* by Leonard and Thomas Digges is indicative of another persistent trend of geographical textbooks from the period, whereby writers cited precedents from ancient sources. Awareness of classical literature was a bedrock of Renaissance educational practices. However, when invoking important figures from antiquity, writers frequently stressed the importance of the use of maps, the commissioning of surveys and scrupulous attention to geographical detail to the success of a multitude of endeavours. Accordingly, young scholars were inculcated with the view that the practical application of geographical knowledge was crucial to achievement in many occupations, particularly those of high status and eminence.

A major feature of this mode of thought was the stress on the militaristic use of maps. In William Shakespeare's *Richard III,* cartography is overtly associated with military planning – before the Battle of Bosworth Field, Richmond announces:

> Give me some ink and paper in my tent
> I'll draw the form and model of our battle,
> Limit each leader to his several charge,
> And part in just proportion our small strength.
> (5.4.21–24)[56]

Richmond reflects a wider recognition of the role of maps in warfare. To emphasise the importance of a knowledge of cartography, educational theorists often cited renowned generals of classical antiquity such as Alexander the Great, attributing their successes to their familiarity with the instruments, processes and products of the discipline of mapmaking. By way of contrast, the military failures of other statesmen and generals were blamed on their ignorance of the geography of the battle site and their unwillingness to utilise maps in preparation for military engagement. Elyot's gesture towards the increasing proliferation of geographical treatises and textbooks and his emphasis on the delights of map study for the 'wytte' were also accompanied by an attentiveness to the socio-political, economic and militaristic potentialities of mapmaking. As well as proceeding from an acknowledgement of the burgeoning cartographic discourse and the potential delight that viewing maps induced in the reader, the encouragement of the use of cartographic objects in education was also motivated by a desire to develop in young scholars a capacity for the practicalities of statecraft, and proficiency in military affairs.

To enact this pedagogical strategy, English textbooks once more drew inspiration from continental writers, most obviously Baldassare Castiglione and Niccolò Machiavelli. In Machiavelli's *The Art of War*, published in 1521, the Florentine thinker writes:

> The first thing he [a military general] ought to do is to get an exact map of the whole country through which he is to march so that he may have a perfect knowledge of all the towns and their distance from each other, and of all the roads, mountains, rivers, woods, swamps, and their particular location and nature.[57]

Castiglione, in his delineation of court politics, *The Book of the Courtier* (1528), foregrounds the importance of painting for the pursuit of 'military purposes':

> In fact, from painting, which is in itself a most worthy and noble art, many useful skills can be derived and not least for military purposes: thus a knowledge of the art gives one the facility to sketch towns, rivers, bridges, citadels, fortresses and similar things, which otherwise cannot be shown even if, with a great deal of effort, the details are memorized.[58]

Like Machiavelli's text, *The Courtier* was one of the most widely read books of the sixteenth century.[59] Both stand as clear precedents for Elyot's own humanist textbook in style, structure, theme and intended audience.[60]

Surveyed for action: The case of Humphrey Gilbert

The emphasis placed by Castiglione and Machiavelli on the military usefulness of maps percolates through many English educational treatises. Alongside a sensitivity to the practical benefits of a familiarisation with maps and globes in an educational setting, and also the predominant belief in the pleasure of map study, Elyot's twin motivations for the 'demontration of cosmographie, by materyall fygures and instruments' was accompanied by texts which made an express point of articulating the utility of maps in the military sphere. In the writings of Humphrey Gilbert, the interest in navigation and the geographical knowledge buttressing such activities epitomised the practical intentions of many English gentlemen interested in the world describing. A renowned explorer and prominent administrator for Elizabethan Ireland, Gilbert's fascination with mapping and its adjunct disciplines is evidenced in *The erection of an achademy in London for educacion of her maiestes wardes, and others the youth of nobility and gentlemen* (c.1573), an educational pamphlet composed in the middle of several notorious campaigns to the colony in the 1560s and 1570s. Gilbert's tract writes Rory Rapple, was 'aimed to secure the education of wards and the younger sons of gentlemen in "matters of accion meet for present practize, both of peace and warre"'.[61] It advances an integration of practical geopolitical aims into the educational mode and includes a description of the model teaching faculty. Tellingly, such a description incorporates a recommendation to employ a number of instructors teaching disciplines overtly relevant to geography. 'Yt were good', writes Gilbert, enumerating his ideal 'Readers', 'Scholemaisters' and 'Vshers':

> That, for their better educacions, there should be an Achademy erected in sorte as followeth: [...] there shalbe placed two Mathematicians, And the one of them shall reade Arithmetick, and the other day Geometry, which shalbe onely employed to Imbattelinges, fortificacions and matters of warre, with the practiz of Artillery and vse of all manner of Instrumentes belonging to the same.[62]

As the reference to 'Imbattelinges, fortificacions and matters of warre, with the practiz of Artillery and vse of all manner of Instrumentes belonging to the same' suggests, underpinning Gilbert's recommendations regarding the idealised 'achademy' was a clear and unequivocal martial motivation. Immediately after this passage, Gilbert discourses on the importance of knowledge of specific subjects, many of which have a militaristic character such as 'Canonrie', 'encampinges' and 'powder and shotte'. Like its continental counterparts, Gilbert's educational paradigm presents practical, and in particular, military enterprise as the primary rationale for teaching the principles of mathematics, geometry and arithmetic through, in part, the practical use of cartographic objects.

This rationale was reiterated by Gilbert in recommendations regarding instruction in the practice of navigation and the production of cartography, in particular, sea charts:

> The other Mathematicians shall reade one day Cosmographie and Astronomy, and the other day the practizes thereof, onely to the arte of Navuigacion, with the knowledge of necessary states making use of Instrumentes appertaining to the same[.] [...] Also there shalbe one who shall teache to drawe mappes, Sea chartes, &c., and to take by view of eye the platte of any thinge, and shall reade the growndes and rules of proportion and necessarie perspective and mensuration belonging to the same, and shalbe yearely allowed.[63]

Unlike Elyot, there is little if anything in Gilbert's representation of the exemplary Elizabethan school institution that gestures towards the pleasure and delight many prominent figures derived from maps and map-reading. However, *Queene Elizabeth's Achademy* manifests the centrality of military endeavour to the promotion of geographical education and principally a familiarity with maps and globes. The emphasis upon the usefulness of the teaching of geography in work intended to inculcate 'serviceable vertues to their prince and Cowntey' in youth he viewed as languishing in 'Idlenes and lascivious and pastimes' reminds us again of the close link between geographic epistemologies and applied methodologies.[64] Typifying the theoretical teaching/practical application model espoused by Gilbert is his advice on how the subject should be taught: 'The other Mathematicians shall reade one day Cosmographie and Astronomy, and the other day the practizes thereof onely to the arte of Navuigacion, with the knowledge of necessary stares making use of Instrumentes appertaining to the same'.[65] The alternating pedagogical structure endorsed by Gilbert – where mathematical principles are taught one day, and the next, the practical endeavours dependent upon such principles are imparted – mirrors the consciously practical imperatives integral in learning practices. As Anthony Grafton and Lisa Jardine have shown, Renaissance education was increasingly imbued with a practical purpose in the face of the unprecedented number of printed materials emerging in the sixteenth century. Citing the likes of Gabriel Harvey and Philip Sidney, Grafton and Jardine write:

> [C]ritical reading, skilful annotation and active appropriation emerge as the central skills, not just of the student of history, but of the intellectual *tout court*. Reading always leads to action - but only proper reading, methodical reading – reading in the manner of a Gabriel Harvey. [...]

Like Harvey, Sidney saw the chief task of the intellectual, ancient or modern, as serving as 'a Discourser, which name I give to who soever speaks not just concerning what happened, but about the qualities and circumstances of what happened [...]' – a definition that embraces both what Harvey saw in Livy and what he hoped himself to become.[66] Grafton and Jardine's model, where 'reading always leads to action' and a focus is retained on 'not just concerning what happened, but about the qualities and circumstances of what happened' was a prominent feature of geographical teaching. As the likes of Gilbert and Machiavelli demonstrate, the promotion of the imparting of knowledge of geography and a familiarity with maps was stimulated by practical concerns and military purposes.

Theory and practice: Reading and surveying together

A text which draws together the two motivations undergirding the advocacy of cartographic usage in the classroom – practical, militaristic use of maps and the delight of cartographic contemplation – is Thomas Blundeville's *A briefe description of vniuersal mappes and cardes*, published in 1589. A prolific and diverse writer, mathematics tutor and companion of figures such as John Dee, Blundeville had interests in navigation, and his writings on this subject 'were principally directed towards young gentlemen, providing instruction on astronomy, maps, and instruments'.[67] In his treatise the author sets out his intention to enhance the geographical knowledge of his readers and also improve their acquaintance with cartographic objects. Observing that many of his contemporaries 'delight to looke on Mappes, and can point to England, France, Germanie, and to the East and West Indies', Blundeville also states that despite this curiosity, many 'yet for want of skill in Geography, [and] they knowe not with what maner of lines they are traced, nor what those lines do signifie, nor yet the true vse of Mappes in deed'.[68]

To correct this ignorance, Blundeville's guide takes the reader through several contemporary maps published by cartographers such as Mercator and Frisius with the intention of familiarising the reader with some of the rudiments of mapping and its uses. In doing so, the author frequently employs an elaborate descriptive method that implies the necessary presence of both text and map during reading for the scholar to fully digest Blundeville's work. Repeatedly, Blundeville points to precise features of the particular cartography discussed, as in the detailed comparison in the use of latitude between ancient maps and their more contemporary 'moderne' successors:

> But the ancient and moderne doe greatly differ in the diuision of the partes of latitude, as well Northerne as Southerne, and also in longitude: for, whereas the ancient Cosmographers doe deuide each latitude into 90. degrees by certaine Paralels making 9. equall spaces, euery space containing 10. equall degrees: in the latter Mappes last mencioned, you shall finde those spaces and the degrees thereof altogether vnequall, the first 3 spaces next the Equinoctiall only excepted, for those differ not aboue one halfe degree at the most: but from thence Northward, euery space is greater then other, and euery degree in euery such space is greater then other, insomuch as the fourth space containeth 11 degrees and a halfe of those degrees which are set downe in the first space, and the fift[h] space conteineth of such degrees 13 degrees 3/4, the 6 space containeth of the said degrees 16 degrees 1/4: ye 7 containeth of the same degrees 20 degrees 1/2, so as the space is is twise so broade as the first space and one halfe degree more: the eight space conteineth of the said first degrees 36: further then which 8 spaces containing 80 degrees of latitude, their Mappes extend not Northward: and they obserue the like proportion in the Southerne latitude, sauing that they extende no farther Southward then to 66 degrees and a halfe.[69]

In this illustrative excerpt, Blundeville sets out the physical characteristics of the map in question – 'in the latter Mappes last mencioned, you shall finde those spaces and the degrees thereof altogether vnequall'. The comparative analysis, the level of cartographic and chart detail, and the complex ratios detailed suggest that the reader would be expected to view Blundeville's text alongside the maps being discussed. The rhetorical gesturing towards the map ('you shall finde', 'you may see'), recurs regularly in Blundeville's treatise, such as where he writes that '*Asia* is bounded on the South with diuers other gulphes & seas, as you may see in the Map'; or 'And first you haue to vnderstand, that the Meridians which you see in the Mappe, doe serue for diuers purposes'.[70] In addition, Blundeville's treatise also encourages the reader to utilise cartographic instruments alongside maps and globes to facilitate the reading and understanding of his work. In a passage advising how to calculate distance between two points, 'if the two places hauing one selfe latitude, doe differ onely in longitude' Blundeville enjoins:

> And if you see that the two places in the mappe doe stand far a sunder, then for the more speedines, take with your compasse fiue such degrees at once, being first prickt vpon a peece of paper which is iust 300 miles, and at the widenes measure the sayd space, and if there remain at the last any od space, then straighten your Compasse and fit them to that odde space, and looke how many of the foresaid degrees that comprehendeth, and hauing multiplied the same by 60 adde the product thereof to the former summe.[71]

The reader is encouraged to use the compass on their own map to calculate the distance between two points and, as a result, learn about maps and their properties by demonstration. By utilising the compass, the cartographic scholars can apply established scalar calibrations and so familiarise themselves with the map reading skill of determining distance and, indeed, the principles that buttress such a skill.

Blundeville's promotion of a practical engagement with maps and globes when learning the principles of navigation is maintained in *His Exercises* (1594), a treatise aimed at the 'furtherance of [the] art of nauigation'. Discussing the toponymic origin of certain geographic locations and features, Blundeville emphasises the importance of using maps and globes when learning about geography:

> Now if you would knowe what Kingdomes, Regions, Cities, Townes, Seas with their Hauens, Ports, Bayes, and Capes, Iles, Floods, Marishes, and Mountaines, are contained in euery one of these foure parts, then peruse often the vniuersall Maps, a[n]d Terrestriall Globes, as well of the moderne as of the auncient Writers, and also the Tables of *Ptolomie* and of *Ortelius,* which I wish that they had beene made in such forme as the Tables of *Ptolomie* are: for hauing the North alwaies set in the front, it should be the readier to compare the shape or situation of any place or Region to the vniuersall Mappe, and by knowing the Longitude and Latitude of any place, it should be the easier to finde the same, as well in the speciall Table as in the vniuersall Mappe or Globe.[72]

Blundeville's instruction to 'peruse often the vniuersall Maps, a[n]d Terrestriall Globes' alongside geographic textbooks both ancient and contemporary ('the moderne as of the auncient Writers, and also the Tables of *Ptolomie* and of *Ortelius*') replicates the rhetorical mode of his earlier treatise. The stress on physical comparison between cartographic depiction and real location ('shape or situation of any place or Region'), the insistence upon the presence of the map alongside the text, and the application of the principles of longitude and latitude to determine both location and distance all serve to emphasise Blundeville's promulgation of a practical engagement with cartography. *His Exercises*, intended as a humanist educational treatise in a similar vein to Elyot's *Boke*, reiterates its predecessor's emphasis upon learning geography by demonstration.

Peacham's pedagogy

The promotion of the use of cartographic objects in geographic education and the support for a practical familiarisation with the process and principles of cartography continued into the seventeenth century in England. An example of the persistence of this tendency can be found in the writings of Henry Peacham. Peacham's exposition on education, *The compleat gentleman fashioning him absolute in the most necessary & commendable qualities concerning minde or bodie that may be required in a noble gentleman* (1622) is one of the most extensive seventeenth-century discussions of the benefits of the usage of cartographic objects within an educational curriculum. Written as 'a final effort to gain royal (or London) patronage' following the death of Prince Henry in 1612 and inspired by Peacham's belief 'that the education of young English gentlemen was markedly inferior to that afforded to European gentry'[73] *The Compleat Gentleman* offers, in the tradition of Elyot and Blundeville, advice regarding 'fashioning him absolute in the most necessary & commendable qualities concerning minde or bodie that may be required in a noble gentleman'.

Cartography, and in particular cartographic objects, feature prominently in this 'fashioning'. The first reference to 'mappes' appears towards the end of chapter six, which discourses on 'stile in speaking and writing, and of Historie'. Peacham concludes the chapter with a brief excursus on the importance of keeping books in good condition – 'suffer them not to lie neglected [...] and goe in torne coates, who must apparell your minde with the ornaments of knowledge, aboue the roabes and riches of the most magnificent Princes' – and describes the ideal physical environment of the learning space to achieve such an aim:

> To auoide the inconuenience of moathes and moldinesse, let your studie be placed, and your windowes open if it may be, towards the East, for where it looketh South or West, the aire being euer Subiect to moisture, moathes are bred and darkishnesse encreased, whereby your mappes and pictures will quickly become pale, loosing their life and colours, or rotting vpon their cloath, or paper, decay past all helpe and recouerie.[74]

In this passage, Peacham draws up a map of the ideal study, complete with compass signifiers. Noting the fluctuations of 'aire' and 'moisture', the author constructs a space for learning that is sensitive to the fragility and tactileness of maps, globes and other cartographic objects. The ideal scholar should, in many respects, attempt to imitate this cognition. Presupposing the presence of 'mappes' as one of the fixtures of the study, Peacham reflects the support for the use of cartographic objects in educational processes and environments. Additionally, Peacham's treatise also pays especial heed to the imagistic character of these objects – allying them to 'pictures', the author emphasises the importance of preserving the quality, clarity and vividness of their visual characteristics. Thus, the location of the study must be predicated on the avoidance of 'moathes [moths]', 'moldinesse' and 'darkishnesse' and the preservation of the 'life and colours' of the maps.

Such a stress on the visual qualities of the map and its preservation can be linked to the author's interests in graphic arts. Peacham's passion for illustration, drawing and the graphic arts was lifelong. In 1595, as a young seventeen-year-old student at Cambridge, he produced what is widely accepted as the first illustration of a Shakespeare play, a drawing of a scene from *Titus Andronicus*. The author's earliest manuscript emblem books were amalgams of text and image, and in 1606 Peacham published a book entitled *The Art of Drawing with the Pen*, a treatise intended 'for the behoofe of all young Gentlemen, or any els that are desirous for to become practicioners in this excellent, and most ingenious Art'.[75] The stress on the need for preservation of maps in *The Compleat Gentleman* reveals Peacham's concern with the pictorial qualities of the cartographic document, with the maintenance of its 'life and colours' being of vital importance to its status in an educational environment such as the study or classroom. Recalling at once the encomiastic tracts penned by such cartographic enthusiasts as John Dee and William Cuningham in its emphasis on the brightness and colouration of maps, Peacham also reiterates obliquely Elyot's assertion of the cartographic object's inherent trait of 'refresshynge the wytte' and as a result providing 'a conuenyent lesson' for the scholar.

Perhaps Peacham's most extended discussion of the usage of cartographic objects is found, unsurprisingly, in a chapter entitled 'Of Cosmographie'. As previously discussed, for educationalists, cosmography was a vast subject, covering diverse scientific disciplines. *The Compleat Gentleman*'s own definition repeats such a broad classification: 'Cosmography', Peacham writes, 'containeth Astronomie, Astrologie, Geography and Chorography'.[76] Preceding this description, however, Peacham also draws attention to the practical use of maps in the pursuit of military leadership. Peacham reminds his reader of the failure of Marcus Licinius Crassus, a Roman general defeated and killed at the Battle of Carrhae in 53BC, and attributes this loss in part to a lack of knowledge: 'And the foule ouerthrow that Crassus receiued by the Parthians, was imputed to nothing else, [t]hen his ignorance of that Countrie, and the passages thereof'.[77]

Providing a stark contrast to such ignorance, however, is Alexander of Macedon, perhaps the most famed general in classical history:

> *Alexander,* therefore taking any enterprise in hand, would first cause an exact mappe of the country to bee drawne in collours, to consider where were the safest entrance, where he might passe this Riuer, how to auoide that Rocke, and in what place most commodiously giue his enemie battaile.[78]

Such a referential framework enables Peacham to set up a binary of military failure and success, reinforcing his stress on the advantages of learning about geography and, specifically, the practical benefits of using maps. The divergent fortunes of Crassus and Alexander stem in part, Peacham asserts, from their varying knowledge of geography. Crassus' defeat at the hands of the Parthians is caused by 'ignorance of that Countrie, and the passages thereof'. Alexander, by comparison, is successful because of his familiarity with the surrounding terrain. Furthermore, coupled with Alexander's superior geographic knowledge is his engagement with maps ('therefore taking any enterprise in hand') and also his active participation in the production of cartography – the great general, unlike Crassus, commissions a 'mappe of the country to bee drawne in collours'. In Peacham's treatise, Crassus serves as a warning of the dangers for the military leader of geographic ignorance, while Alexander, by contrast, exemplifies the benefits of map reading and map production in the military arena.

Into the 1600s: Gailhard, Milton and Bacon

The Compleat Gentleman is one of many seventeenth-century tracts that maintain the convention of Elyot regarding the 'demontration of cosmographie, by materyall fygures and instrumentes'. In *The compleat gentleman, or, Directions for the education of youth* (1678) the French writer and religious controversialist Jean Gailhard recommends to the young scholar that:

> it will be well to have before you the Maps of every Province, and, if possible, of the Towns you are in, to know the right situation thereof (which also may be done by getting upon some Steeple, or high place) and learn their Frontiers and Neighbours.[79]

Apparent in the suggestion that 'it will be well to have before you the Maps of every Province' are traces of the cartographic enthusiast's conceptual understanding of what maps could do, in particular, their stress upon the map's transcendental capability. Just as Robert Herrick's brother may in his 'map securely sail' and view the world within the confines of his country house, so the act of map reading represents for the scholar a ready substitute for 'getting upon some Steeple, or high place'. Gailhard, much like Peacham, promoted the map as an apparatus of learning. It imparts information regarding 'the right situation' of 'every Province', enabling the student to gain knowledge of their boundaries and bordering countries ('Frontiers and Neighbours'). Gailhard maintains his recommendation of cartographic objects as instruments of education later on in his treatise, writing:

> The use of the Map will be very beneficial if he understands it, which he can do easily; this will give a great light to some parts of History, depending upon Geography[.] [...] [H]e must endeavour to understand the use of the Terrestrial Globe, which can much help him therein.[80]

Like Peacham, Gailhard stresses the multi-disciplinary application of cartographic objects. Furthermore, the comprehensiveness of cosmography as a subject permeates Gailhard's recommendation of the 'use of the Terrestrial Globe' and the omnipresence of the map during scholarly learning.

However, among many other seventeenth-century educationalists and thinkers who endorse a similar approach, the most renowned is perhaps John Milton. In his thesis *Of Education*, Milton recommends:

> Ere halfe these Authors be read, which will soon be with plying hard, and dayly, they cannot choose but be masters of any ordinary prose. So that it will be then seasonable for them to learn in any modern Author, the use of the Globes, and all the maps first with the old names; and then with the new: or they might be then capable to read any compendious method of naturall Philosophy. And at the same time might be entring into the Greek tongue, after the same manner as was before prescrib'd in the Latin; whereby the difficulties of Grammar being soon overcome, all the Historicall Physiology of *Aristotle* and *Theophrastus* are open before them, and as I may say, under contribution.[81]

Published in 1644, *Of Education* is noteworthy because of the intellectual stature of its author and also because it is addressed to Samuel Hartlib, the prominent educational reformer. Hartlib had sought Milton's views on the subject, and the sparse bibliographic characteristics of the treatise – relative shortness, absence of a title page, author's or publisher's name and publication date – suggest that the author intended it for limited circulation.[82]

However, such a restricted dissemination must be balanced against the audience for whom the treatise was proposed. As a correspondent of Hartlib's, Milton was involved in the so-called 'Hartlib Circle', one of the most influential pedagogical forums of the seventeenth century.[83] In his editorial introduction to the treatise, Rosenblatt foregrounds Milton's stress on the need for a practical engagement with certain recognised disciplines. Much like Elyot, Milton proposes a familiarity with the objects used in the practice of different disciplines, including mapmaking and map reading. Citing the ideas of the Continental educationalist John Amos Comenius, whose theories influenced Hartlib, Rosenblatt locates *Of Education* within the wider context of early modern educational discourse, foregrounding of the use of instruments in the learning process:

> The treatise is sympathetic to the practical nature of a Comenian education, which was influenced by Francis Bacon's scientific empiricism and inductive logic. The *organum* in Bacon's influential *Novum Organum* (1620) translates as instrument or implement, used of military or architectonic engines, and Milton conceives of logic and rhetoric as 'organic arts,' serving as means to an end: 'language is but the instrument conveying to us things useful to be known.' He advocates supplementing book learning with the 'helpful experiences of hunters, fowlers, fishermen, shepherds, gardeners, apothecaries; and in other sciences, architects, engineers, mariners, anatomists.'[84]

In the sixteenth century, maps, globes and cartographic instruments were understood as devices that augmented the educational experience of young children. This belief was linked to an extent to the fascination with cartographic objects evinced by many of the period's most esteemed intellectuals. However, the educational advancement of cartographic objects in the classroom found its basis not only in the cartographic enthusiasm but also in the perceived role cartography played in the execution of successful military endeavour. Reiterating continental precedents, many writers such as Humphrey Gilbert, while largely inattentive to the visual delights of the card or map, focused on the sheer practicality of geography and emphasised the benefits of its practical application. Such a trend was maintained into the seventeenth century by the likes of Henry Peacham and Jean Gailhard. As Jason P. Rosenblatt suggests, this established perception of geography's theory/practice nexus would go on to inform embryonic ideas of inquiry and 'inductive logic'. As we shall see in the next chapter, the art of world describing would play a small but crucial role in the educational theories of perhaps the most influential early modern English pedagogical theorist of them all, Francis Bacon.

Notes

1. Desiderius Erasmus, *The Correspondence of Erasmus: Letters 594–841 (1517–1518)* (Toronto: University of Toronto Press, 1979), p. 275.
2. Jonathan M. Smith, 'State Formation, Geography, and a Gentleman's Education', *Geographical Review*, 86:1 (1996), p. 94.
3. Understandably, the emergence of the printing press has proved a fecund topic of Renaissance historiography. Among a wealth of studies exploring this subject and its monumental impact on European intellectual, cultural and religious life are Marshall McLuhan, *The Gutenberg Galaxy: The Making of Typographic Man* (Toronto: University of Toronto Press, 1962); Elizabeth Eisenstein, *The Printing Press as an Agent of Change* (Cambridge: Cambridge

University Press, 1980); and *The Renaissance Computer: Knowledge Technology in the First Age of Print*, eds. Jonathan Sawday and Neil Rhodes (London: Routledge, 2000).

4. J.N.L. Baker, 'Academic Geography in the Seventeenth and Eighteenth Centuries', *The Scottish Geographical Magazine*, 51:3 (1935), p. 1.

5. As Norman J.W. Thrower remarks, the influence of Ptolemy's text on Western maps and atlases 'would be difficult to exaggerate'. Thrower, *Maps and Civilization: Cartography in Culture and Society*, 3rd edition (Chicago: University of Chicago Press, 2008), p. 58. See also E.G.R. Taylor, *Tudor Geography: 1485–1583* (Oxford: Clarendon Press, 1930) and *Late Tudor and Early Stuart Geography* (Oxford: Clarendon Press, 1934). Scholars such as David N. Livingstone, David Turnbull, Lesley B. Cormack, Charles Withers and Robert Mayhew have examined the various influences of the re-emergence of classical texts on geography in early modern England. See David N. Livingstone, 'Geography, Tradition and the Scientific Revolution: An Interpretative Essay', *Transactions of the Institute of British Geographers*, 15:3 (1990), pp. 359–373; David Turnbull, 'Cartography and Science in Early Modern Europe: Mapping the Construction of Knowledge Spaces', *Imago Mundi*, 48 (1996), pp. 5–24; Lesley B. Cormack, *Charting an Empire: Geography at the English Universities, 1580–1620* (Chicago: Chicago University Press, 1997), especially pp. 17–47 and Charles Withers and Robert Mayhew, 'Rethinking "Disciplinary" History: Geography in British Universities, *c. 1580–1887*', *Transactions of the Institute of British Geographers*, 27 (2002), pp. 11–29. In her essay, 'Map Ownership in Sixteenth-Century Cambridge: The Evidence of Probate Inventories', *Imago Mundi*, 47 (1995), pp. 67–93, Catherine Delano Smith has provided much historical detail regarding the ownership of cartographic textbooks and paraphernalia such as globes and maps among students at Oxford and Cambridge.

6. S.K. Heninger, 'Tudor Literature of the Physical Sciences' in *Huntington Library Quarterly*, 32:2 (1969), p. 118.

7. Heninger, 'Tudor Literature of the Physical Sciences', p. 119.

8. Cormack, *Charting an Empire*, p. 37.

9. *The Invention of Discovery, 1500–1700*, ed. James Dougal Fleming (Surrey: Ashgate, 2011), p. 186.

10. *The Invention of Discovery*, p. 186.

11. Withers and Mayhew, 'Rethinking "Disciplinary" History', p. 13.

12. As Withers and Mayhew observe:

> The modern university formations in which most academics today operate in Europe and North America (with their concomitant conception of disciplines) are those which were codified in the century after 1850, as the humanist reworking of the Aristotelian curriculum which had served the universities since their formation broke down. The modern conception of a university discipline, and therefore of what it means for geography to be a presence in the universities, has usually involved several criteria: a separate degree scheme run in an identifiable and autonomous department, academics trained in that subject and researching an aspect of it and students receiving formal qualifications in that subject.

See Withers and Mayhew, 'Rethinking "Disciplinary" History', p. 12.

13. Withers and Mayhew, 'Rethinking "Disciplinary" History', p. 13.

14. Withers and Mayhew, 'Rethinking "Disciplinary" History', p. 13.

15. For an example of humanist education and its approach, see Montaigne's 'On Schoolmaster's Learning' and 'On Educating Children', Michel de Montaigne, *The Complete Essays*, ed. M.A. Screech (London: Penguin

Classics, 2003), pp. 150–162 and pp. 163–199, respectively; and Bacon's 'Of Studies', Francis Bacon, *The Essays*, ed. John Pitcher (London: Penguin Books, 1985), pp. 209–210. In this latter essay in particular Bacon, with characteristic eloquence, alludes to the comprehensiveness of Renaissance learning: 'Reading maketh a full man; conference a ready man; and writing an exact man. And therefore, if a man write little, he had need have a great memory; if he confer little, he had need have a present wit; and if he read little, he had need have much cunning, to seem to know that he doth not. Histories make men wise, poets witty, the mathematics subtle, natural philosophy deep, moral grave, logic and rhetoric able to contend' (p. 209).

16. William Pemble, *A Briefe Introduction to Geography* (Oxford: John Lichfield, 1630), p. 1. Pemble's treatise was highly popular among seventeenth-century readers and was re-published several times, reaching a fifth edition by 1675.

17. An indicative description of the voluminousness of natural philosophy may be found in the works of one of the most esteemed scientists of the age, the chemist, physicist, inventor and founding member of the Royal Society Robert Boyle. In his textbook *Some considerations touching the vsefulnesse of experimental naturall philosophy* (Oxford: Henry Hall, 1663), Boyle writes: 'For most other Sciences, at least as they are wont to be taught, are so narrow and so circumscrib'd, that he who has read one of the best and recentest Systems of them, shall find little in the other Books publisht on those subjects, but disguis'd repetitions; and a diligent Scholar may in no long time learn as much as the Professors themselves can teach him. But the objects of Naturall Philosophy, being as many as the Laws and Works of Nature are, so various and so numberlesse, that if a Man had the Age of *Methuselah* to spend, he might sooner want time then matter, for his Contemplations: And so pregnant is each of that vast multitude of Creatures, that make up the Naturalists Theme, with usefull matter to employ Mens studie, that I dare say, that the whole life of a Philosopher spent in that alone, would be too short to give a full and perfect account of the Natural Properties and Uses of any one of several Minerals, Plants, or Animals, that I could name' (p. 12).

18. Representative of the commodiousness of 'geographical knowledge' in our time is the current syllabus of the Department of Geography at Marlowe's university Cambridge, whose research areas in 2021 include (among many other subjects) demography, ecology, environmental studies, geology and botany.

19. Withers and Mayhew, 'Rethinking "Disciplinary" History', p. 14.

20. Nathaniel Carpenter, *Geographie delineated forth in two bookes* (Oxford: John Lichfield, 1635), p. 1.

21. Robert J. Mayhew, *Enlightenment Geography: The Political Languages of British Geography, 1650–1850* (Basingstoke: Palgrave Macmillan, 2000), p. 27.

22. Robert Burton, *The Anatomy of Melancholy* (Oxford : John Lichfield & James Short for Henry Cripps, 1621), p. 327. Burton's description of Kepler's work is repeated by the writer Robert Heath in his *Paradoxical Assertions and Philosophical Problems Full of Delight and Recreation for All Ladies and Youthful Fancies* (London: 1659): 'Well may we then admit of *Brunus* infinite Worlds and Suns; of *Keplar's* Lunar Geography; and believe with *Campanella, Galilaeus, Origanus, &c.* that the Earth hath motion, is a Planet, and shines like the Moon, to these Lunar Inhabitants' (p. 10).

23. Baker, 'Academic Geography in the Seventeenth and Eighteenth Centuries', p. 132.

24. Baker, 'Academic Geography in the Seventeenth and Eighteenth Centuries', p. 132.

25. Cormack, *Charting an Empire*, p. 59.

26. John Dee, *The elements of geometrie of the most auncient philosopher Euclide of Megara* (London: John Day, 1570), sig B3r.

27. Cormack, *Charting an Empire*, p. 163.
28. Lucia Nuti, 'Mapping Places: Chorography and Vision in the Renaissance', *Mappings*, ed. Denis Cosgrove (London: Reaktion Books, 2000), p. 90.
29. For a detailed analysis of chorography in early modern England, see Stan Mendyk, 'Early British Chorography', *The Sixteenth Century Journal* 17:4 (1986), pp. 459–481; Monica Matei-Chesnoiu, *Early Modern Drama and the Eastern European Elsewhere: Representations of Liminal Locality in Shakespeare and His Contemporaries* (New Jersey: Associated University Press, 2009), pp. 66–68; and Rhonda Lemke Sanford, *Maps and Memory in Early Modern England: A Sense of Place* (Basingstoke: Palgrave Macmillan, 2002), p. 147, n. 22. Sanford's analysis explores the status of chorography in the hierarchy of geographic disciplines, citing Ptolemy's description, laid out in his *Geography* that 'The end of Chorography is to deal with a part of the whole, as if one were to paint only the eye or the ear itself. The task of Geography is to survey the whole in its just proportions, as one would the entire head' (p. 147, n. 22). Elsewhere, Barbara J. Shapiro discusses the discipline in the context of the economic politics of seventeenth-century Britain and 'Baconian and Royal Society-sponsored "histories of trade" that charted developments in trade, manufacturing and agriculture'. Barbara J. Shapiro, *A Culture of Fact: England, 1550–1720* (Ithaca: Cornell University Press, 2003), p. 66.
30. John Speed, *The Theatre of the Empire of Great Britaine* (London: William Hall, 1611), title page.
31. Dee, *Euclide*, sig. A4r.
32. Henry Peacham, *The Compleat Gentleman* (London: John Legat, 1622), p. 58.
33. David Fletcher, 'Saxton, Christopher (1542x4–1610/11)', *ODNB*, (Oxford: Oxford University Press, 2004), online edn. January 2008, http://www.oxforddnb.com/view/article/24760 (accessed 3rd July 2018). For an examination of Saxton's work as a surveyor, his role in the national survey of the 1570s and his cartographic output, see Sarah Tyacke and John Huddy, *Christopher Saxton and Tudor Map-Making* (London, British Library Reference Division, 1980).
34. The 1586 edition of Camden's text was in Latin. The original title reads: 'Britannia siue Florentissimorum regnorum, Angliae, Scotiae, Hiberniae, et insularum adiacentium ex intima antiquitate chorographica descriptio'. Camden, *Britannia* (London: Ralph Newbery, 1586), title page.
35. William Cuningham, *The Cosmographical Glasse* (London: John Day, 1559), title page.
36. Cormack, *Charting an Empire*, p. 622.
37. Lesley B. Cormack, 'Maps as Educational Tools in the Renaissance', *The History of Cartography, Volume Three, Part 1: Cartography in the European Renaissance*, ed. David Woodward (Chicago: Chicago University Press, 2007), p. 625.
38. Elly Dekker, 'Globes in Renaissance Europe', *The History of Cartography, Volume Three, Part 1: Cartography in the European Renaissance*, ed. David Woodward (Chicago: Chicago University Press, 2007), pp. 149–150.
39. Dekker, 'Globes in Renaissance Europe', p. 150. Dekker describes Apian's model: 'This paper instrument consisted of a base plate, a volvelle with a printed map of the earth, and another movable part shaped as the rete of an astrolabe. In addition, around the north pole, there is a smaller hour circle with an index arm and another index arm with a latitude scale, and both can rotate around the centre' (p. 150).
40. Notably, Apian's later work *Cosmographia Introductio* ['Introduction to Cosmography'] (1529) was listed included in the book lists of several colleges at Oxford and Cambridge, so the author would have been familiar to English students in the sixteenth century. See Cormack, 'Maps as Educational Tools in the Renaissance', pp. 107–112.

41. Cormack, *Charting an Empire*, p. 26.
42. Cormack, *Charting an Empire*, pp. 26–27.
43. Henry Billingsley in Dee, *Euclid*, A2r.
44. Leonard and Thomas Digges, *A Geometrical Practical Treatize Named Pantometria* (London: Abell Jeffes, 1571), sig. A2r. Nathaniel Carpenter praised the contribution of Leonard Digges' work to the geographical geometry in *Geography Delineated Forth in Two Bookes* (1635). For a discussion of Digges' extensive involvement in the development of the subject in later sixteenth-century England, see A.W. Richeson, *English Land Measuring to 1800: Instruments and Practices* (Cambridge, Massachusetts: Society for the History of Technology and MIT Press, 1966), pp. 52–57; Joy B. Easton, 'Leonard Digges', *Dictionary of Scientific Biography*, Volume 2, ed. C.C. Gillispie (New York: Charles Scribner's Sons, 1980), p. 97.
45. William Poole, 'Early Modern Eclecticism', *Critical Quarterly*, 52:4 (2010), p. 12.
46. For a detailed analysis of this growing scholarly population, see Lawrence Stone, 'The Educational Revolution in England, 1560–1640', *Past & Present*, 28:1 (1964), pp. 41–80. '[M]en with no interest in a clerical career', notes Lesley B. Cormack, 'began to see a few years at Oxford or Cambridge as a valuable part of their practical education, aiding them in future careers as merchants, politicians, courtiers or even country gentleman'. Cormack, *Charting an Empire*, p. 23.
47. Smith, 'State Formation, Geography, and a Gentleman's Education', p. 91.
48. Cormack, 'Maps as Educational Tools in the Renaissance', p. 622.
49. The three educational treatises that form the main focus of this chapter were immensely successful. Elyot's *Boke* was published in nine editions between 1531 and 1580. Similarly, Blundeville's *Exercises* was published seven times between 1594 and 1638. Peacham's *Compleat Gentleman*, meanwhile went through four separate editions between 1622 and 1661.
50. Cormack, 'Maps as Educational Tools in the Renaissance', p. 624.
51. See in particular Gavin Hollis, '"Give Me the Map There": King Lear and Cartographic Literacy in Early Modern England', *Portolan*, 68 (2007) pp. 8–25.
52. Thomas Elyot, *The Boke Named the Governour* (London: Thomas Berthelet, 1531), p. 35.
53. E.G.R. Taylor's exhaustive catalogue lists 159 'geographical or kindred works' published in the 49 years between the first (1531) and ninth (1580) editions of Elyot's treatise. Taylor, *Tudor Geography: 1485–1583*, pp. 168–173.
54. Elyot, *The Boke Named the Governour*, p. 35.
55. Elyot, *The Boke Named the Governour*, pp. 23–24.
56. All quotations are taken from William Shakespeare, *The Oxford Shakespeare: The Complete Works*, eds. John Jowett *et al.*, 2nd edition (Oxford: Oxford University Press, 2005).
57. Niccolò Machiavelli quoted in Arthur F. Kinney, *Shakespeare's Webs: Networks of Meaning in Renaissance Drama* (New York: Routledge, 2004), p. 111.
58. Baldassare Castiglioni, *The Book of the Courtier* (London: Penguin, 1976), p. 97.
59. The Elizabethan courtier Thomas Hoby first translated *Il Cortegiano* into English in 1561. L.G. Kelly describes the influence of Castiglione's book on England in the period after its publication: 'Setting the standards of social behaviour for the English cultivated public, *The Courtier* [also] left a profound mark on Elizabethan literary and stylistic practice'. L.G. Kelly, 'Hoby, Sir Thomas (1530–1566)', *ODNB* (Oxford University Press, 2004), online edn. January 2008,http://www.oxforddnb.com/view/article/13414 (accessed 4th July 2014).

60. David Starkey notes Elyot indebtedness to its continental forebears. Starkey, 'The Court: Castiglione's Ideal and Tudor Reality; Being a Discussion of Sir Thomas Wyatt's Satire Addressed to Sir Francis Bryan', *Journal of the Warburg and Courtauld Institutes*, 45 (1982), p. 233.
61. Rory Rapple, 'Gilbert, Sir Humphrey (1537–1583)', *ODNB*, http://www.oxforddnb.com/view/article/10690 (accessed 7th July 2014).
62. Humphrey Gilbert, *Queene Elizabethes Achademy*, ed. F.J. Furnivall (London: Early English Text Society, 1869), pp. 1–4.
63. Gilbert, *Queene Elizabethes Achademy*, p. 5.
64. Gilbert, *Queene Elizabethes Achademy*, p. 1.
65. Gilbert, *Queene Elizabethes Achademy*, p. 5.
66. Anthony Grafton and Lisa Jardine, 'Studied for Action: How Gabriel Harvey Read His Livy', *Past and Present*, 129 (1990), pp. 76–77.
67. Tessa Beverley, 'Blundeville, Thomas (1522?–1606?)', *ODNB* (Oxford: Oxford University Press, 2004), online edn. January 2008, http://www.oxforddnb.com/view/article/2718 (accessed 7th July 2014).
68. Thomas Blundeville, *A briefe description of vniuersal mappes and cardes* (London: Roger Ward, 1589), A3r.
69. Blundeville, *A briefe description of vniuersal mappes and cardes*, B4r.
70. Blundeville, *A briefe description of vniuersal mappes and cardes*, C4r.
71. Blundeville, *A briefe description of vniuersal mappes and cardes*, D4v.
72. Thomas Blundeville, *M. Blundevile his exercises containing sixe treatises* (London: John Windet, 1594), p. 300.
73. F.J. Levy, 'Henry Peacham and the Art of Drawing', *Journal of the Warburg and Courtauld Institutes*, 37 (1974), p. 177. John Horden, 'Peacham, Henry (b. 1578, d. in or after 1644)', *ODNB* (Oxford University Press, 2004), online edn. January 2008, http://www.oxforddnb.com/view/article/21667 (accessed 7th July 2019).
74. Peacham, *The Compleat Gentleman*, p. 55.
75. For a discussion of this text and its significance in English art history, especially in relation to the burgeoning interest in non-portraiture art in the early seventeenth century, see F.J. Levy, 'Henry Peacham and *The Art of Drawing*', pp. 174–190; and L.E. Semler, 'Breaking the Ice to Invention: Henry Peacham's *The Art of Drawing* (1606)', *The Sixteenth Century Journal*, 35:3 (2004), pp. 735–750. Semler foregrounds Peacham's influence on the development of aesthetic taste in the period, acclaiming *The Art of Drawing* for 'introducing English readers to continental art theory and pedagogy' (p. 735).
76. Peacham, *The Compleat Gentleman*, p. 58.
77. Peacham, *The Compleat Gentleman*, p. 57.
78. Peacham, *The Compleat Gentleman*, p. 57.
79. Jean Gailhard, *The Compleat Gentleman, or, Directions for the Education of Youth* (London: Thomas Newcomb, 1678), p. 37.
80. Jean Gailhard, *The Compleat Gentleman*, p. 54.
81. John Milton, *Of Education* (London: 1644), p. 4.
82. John Milton, *Milton's Selected Poetry and Prose*, ed. Jason P. Rosenblatt (New York: W. W. Norton and Company, 2011), p. 318.
83. For a more extensive analysis of the Hartlib Circle's influential role in seventeenth-century intellectual thought, see *Samuel Hartlib and Universal Reformation: Studies in Intellectual Communication*, eds. Mark Greengrass, Michael Leslie and Timothy Raylor (Cambridge: Cambridge University Press, 1994). The online database *Hartlib Circle* (http://www.mhs.ox.ac.uk/gatt/page_index.php?section=circle (accessed 7th July 2014)), set up by the University of Oxford, provides a useful introduction to the group, alongside a brief, though the illustrative selection of their writings.
84. Milton, *Milton's Selected Poetry and Prose*, p. 319.

5 Francis Bacon and geographic science

What did Francis Bacon make of 'mapp'ry', that evocative Shakespearean neologism connoting maps, map-making and map-reading?[1] Renowned as a deep thinker on the nature and purpose of scientific endeavour, and regularly identified as a pivotal figure in early modern intellectualism, he possessed, according to Brian Vickers, 'a powerful intellectual grasp, analytical penetration, a mastery of the expressive resources of language [and] the ability to adapt style to subject-matter and purpose'.[2]

We might therefore expect Bacon to be deeply engaged with contemporary mapping activities, mappers and the knowledge systems sustaining both. As we have seen, cartography and its associated disciplines fascinated many prominent Renaissance intellectuals, from John Dee to John Donne. It was adopted and adapted for a range of purposes by writers such as Speed, Drayton and Spenser, evidencing its fertile diegetic modes. Moreover, Baconianism as a typology of thought emphasising the importance of practical and theoretical approaches to knowledge accumulation is conspicuously consonant with the empirical aspects of geographical science. As Antonia McLean writes:

> Many of Bacon's ideas were derivative, and he cannot be taken seriously as an experimental scientist. But as a thinker he grasped more clearly than any of his contemporaries the nature and magnitude of the changes brought about by new technologies and by the discovery of the New World. He saw immediately that it was through these discoveries and inventions that the whole of European civilization had been altered, the work of craftsmen rather than the ideas of the philosophers had brought in the new age.[3]

'Discoveries' – of new worlds and forgotten geographical texts – and 'inventions' – of projection methods and increasingly sophisticated cartographic instruments – lay at the heart of the geographic revolution in early modern England, driving both the renewed interest in the subject and its innovations. An identifiable Baconian 'mapp'ry' substantiated by extensive

DOI: 10.4324/9781003200376-8

discussion of its practice and purpose would be wholly consistent with the seventeenth-century thinker's oeuvre.

It is perhaps surprising then that Bacon, sometime lawyer, Solicitor-General, Attorney-General, Lord Keeper, Lord Chancellor, Baron Verulam, Viscount St. Alban and putative pioneer of food refrigeration, does not appear to have been much of a cartographic enthusiast. He could not, in contrast to many of his contemporaries, be counted among those of whom John Dee famously remarked that 'some, for one purpose: and some, for another, liketh, loueth, getteth, and vseth Mappes, Chartes, & Geographicall Globes'.[4] Despite occupying prominent positions in both the Elizabethan and Jacobean courts, spheres populated with such avid map-readers as William Cecil, Francis Walsingham and the Sidneys, Henry and Philip, Bacon's engagement with cartography appears, by comparison, slight. He never (as far as we know) drew or commissioned any maps himself. Occasionally, Bacon does reference maps and their purpose in his writings: in his essay 'Of Travel', the writer draws attention to the usefulness of geographical knowledge for civic employment. In doing so, he reflects some of the ideas explored in the earlier examination of English pedagogy's engagement with geography in the early modern period.[5] Additionally, Bacon's considerations on 'the True Greatness of Kingdoms and Estates' testify to his sensitivity to the municipal usefulness of cartography, noting that 'cards and maps' help to illustrate 'the number and greatnesse of cities and townes'.[6]

However, despite a diverse and wide-ranging corpus of work that examined issues as miscellaneous as ecclesiastical disputes, English imperial policy, and legislation regarding duels, geographical science does not appear to have stimulated Bacon to any great extent.[7] Or, at the very least, it never served for Bacon, unlike a whole host of other topics, as a subject for any extensive cogitation. A partial though by no means definitive gauge of this lack of enthusiasm for geography is the scarcity of the specific toponym 'geography' in the compendious Baconian canon.[8] We could caveat such an absence with the acknowledgement that, as outlined in the previous chapter, many early modern intellectuals used terms such as 'cosmography', 'geography', 'astronomy' and 'astrology' interchangeably. Furthermore, it might be said that Bacon does the same: in *The Advancement of Learning,* for example, Bacon extols the virtues of scriptural study, observing:

> So likewise in that excellent Book of Job, if it be resolved with diligence, it will be found pregnant, and swelling with natural philosophy; as for example, cosmography, and the roundness of the world [...] wherein the pensilenesse of the earth, the pole of the north, and the finiteness, or convexity of heaven are manifestly touched.[9]

'Cosmographie', in a formulation analogous to many early modern intellectuals, falls under the wider topic of 'naturall Philosophie'. Bacon, like

John Dee and later Henry Peacham, subsumes geographical detail – the physical attributes or 'roundness' of the globe, its topographical makeup ('pensileness of the earth'), the specification of location ('the pole of the north') – under the rubric of 'cosmographie', associating such facts with conspicuously astronomical information like 'the finiteness or convexity of heaven'. Thus, Bacon may be alluding to the favoured skill of John Speed when, in his articulated utopia of the *New Atlantis,* he describes an area in Salomon's House: 'We also have a mathematical house, where are represented all instruments, as well of geometry as astronomy, exquisitely made'.[10]

Nonetheless, the pointed lack of specificity in Bacon's fleeting references to the act of world describing is anomalous in the context of a corpus of work concerned largely with the scientific endeavour. This vagueness in language is even more noteworthy when we consider that it is entirely uncharacteristic, for Bacon is considered one of the pre-eminent classifiers in the history of science.[11] However, where Bacon does engage with geography and its ancillary disciplines, he bypasses simply classifying what the subject involves or indeed expressing an admiration for maps to consider the much more fundamental concept of how the subject as a science functioned. In doing so, Bacon mapped out an area of knowledge inquiry that would prove representational for his own ideal way of doing science, one which relied on the fundamental premise of a cooperation between theory and practice.

Baconianism surveyed

For many scholars of early modern history, Bacon is a fascinating and simultaneously challenging figure. 'From Voltaire and Blake to Horkheimer and Adorno', summarises Alvin Snider, 'Bacon's work has retained a surprising capacity to polarize opinion'.[12] His very first biography, written by the literary editor and clergyman William Rawley opens an account of Bacon's life with this grandiose introduction: 'Francis Bacon, the Glory, of his Age, and Nation; The Adorner, and Ornament, of Learning; Was born, in York House, or York Place, in the Strand; On the 22nd Day of January; In the Year of our Lord, 1560'.[13] Since Rawley, Bacon's status in intellectual thought as both the 'adorner' and also the 'ornament' of learning has been the subject of varying interpretations as readers have sought to pin down Baconian ideas. Markku Peltonen provides an eloquent summary of the variegated critical reception of Bacon's work:

> Bacon has always had a central but controversial place in historical accounts of early modern philosophy. For some, he was the first spokesman of modern science in general and the father of the inductive method in particular. For others, he was an immoral charlatan who had nothing original to say.[14]

The extent of these disputes and the controversy provoked by attempts to characterise Bacon as a founder of the 'new science' of the later seventeenth century have been attributed to a lack of accurate definition, or rather 'the confusion arising from the different sense in which the term "Baconianism" has been used in European cultural history since the seventeenth century'.[15] Thus, proposes Antonio Pérez-Ramos, 'Baconianism, as a historiographical category, shares the fate of Aristotelianism, Platonism, or even Marxism: it is unavoidable but too polysemic and diffuse'.[16]

Yet while polysemy and diffusion pervade readings of Bacon's ideas, it is clear that a persistent impulse running through his work is the promotion of what Peltonen eloquently terms an 'active or operative science'.[17] At the heart of Bacon's 'active or operative science' is a foregrounding of 'experience': a term Bacon uses to signify pragmatic engagement with the stuff of knowledge and its practical enquiries. As he writes in 'Of Studies', 'experience' and 'studies' should exist in a relationship of epistemological interchange:

> To spend too much time in studies is sloth; to use them too much for ornament, is affectation; to make judgment wholly by their rules, is the humour of a scholar. They perfect nature and are perfected by experience: for natural abilities are like natural plants, that need pruning, by study; and studies themselves do give forth directions too much at large, except they be bounded in by experience.[18]

Experience or practical experimentation and observation help to 'perfect' bookish learning. It gives an evidentiary basis for knowledge that would otherwise appear affected or flimsy. Furthermore, the organic simile used by Bacon indicates the dynamism of this synergy: knowledge accumulation, much like a plant, requires both encouragements to grow and careful 'pruning' to ensure it does not lose focus.

These guiding principles of Baconian science – the combination of theory and exercise in scientific enquiry, and the emphasis on a necessary familiarisation with instruments relevant to the discipline studied – are perceptible in much of Bacon's writing. An early indication of the 'active or operative science' can be found in a letter addressed to the scholar Henry Savile.[19] Written between 1596 and 1602, this epistle discusses 'helps for the intellectual powers' and demonstrates the central role 'exercise' plays in Baconian scientific enquiry. Observing that while 'the discourses of philosophers' explore 'the framing and seasoning of youth to moral virtues, tolerance of labours, continency from pleasures, obedience, honour and the like', Bacon points out that they ignore 'the improvement and helping of the intellectual powers, as of conceit, memory, and judgment'.[20] The author, in response to the apparent absence in the 'discourses of philosophers', sets out his own programme for learning, one which stresses practical experimentation.

'The intellectual powers have few means to work upon them than the will or body of man', Bacon writes, 'but one prevaileth, that is exercise, worketh more forcibly in them than the rest'.[21] The encouragement of 'exercise' – with its functional connotations – is developed further and given rationale:

> [E]xercises are to be framed to the life; that is to say, to work ability in that kind, whereof a man in the course of actions shall have most use. [...] The marshalling and sequel of sciences and practices: Logic and Rhetoric should be used to be read after Poesy, History, and Philosophy. First exercise to do things well and clean; after promptly and readily.[22]

In editing Bacon's letter, Brian Vickers has noted that '[i]t is another example of Bacon using the genre of advice literature to communicate his thinking on specific topics'. Furthermore, 'many of the ideas on mankind's learning capacity sketched here are worked out more fully in the *Essays* and the *Advancement of Learning*'.[23] Vickers' summary signposts how the ideas proposed in Bacon's most substantial work of epistemological theory germinated even in his early career as a writer.

The emphasis on the implementation of 'direct exercises' in the didactic environment and the 'framing' of such exercises to the 'life' or pragmatic realities of certain disciplines in his letter to Savile shows that Bacon's ideas of scientific enquiry foregrounded experiential learning. As a corollary, the importance of the use of experiments and instruments in the 'advancement of learning' is also emphasised. The stress on the need for applied experience of the processes and procedures of a specific discipline is apparent most clearly in Bacon's axiom that 'if a man exercise shooting, he shall not only shoot nearer the mark but draw a stronger bow'.[24] Exercise increases astuteness and acuity ('he shall shoot nearer the mark') and also enhances the skilled use of the instrument ('but draw a stronger bow'). 'Operative and contemplative knowledge', asserts Lisa Jardine, 'are [for Bacon] faces on one coin, bound together under philosophy as alternative aspects of a rational understanding of nature'.[25] As the advisory epistle to Savile demonstrates, even before its more refined articulations in works such as *The Advancement of Learning* and the *Novum Organum*, Baconian thought conceived of science as both a meditative pursuit and simultaneously a practical endeavour and 'exercise'.

Geography as a Baconian prototype

Such an approach coincides with geographical enquiry, suggesting that Bacon's ideas of scientific inquiry, to some extent, found their prototype in the nascent cartographic consciousness of the age. Significantly, Lesley Cormack has documented the foreshadowing of Baconian traits in certain sixteenth-century geographical texts. Discussing Richard Hakluyt's *Principall Navigations* (1589–1600), Cormack writes:

> Hakluyt's book [...] reveals an interesting movement towards a 'Baconian' or collecting methodology in the human and descriptive sciences. The collection of useful facts – insignificant when taken individually but amalgamated into a complete worldview that was greater than the sum of its parts – correspond to Francis Bacon's model of tabulation.[26]

Bacon's penchant for classification and enumeration can be similarly identified in the paeans to map study of Elyot, Blundeville and Dee. Likewise, Hakluyt's travel guide anticipated another Baconian principle – the use of instruments in learning. In the preface to the first edition of 1589, Hakluyt announces the latest innovation in English cartographic instruments, Emery Molyneux's terrestrial globe, the first of its kind made in England and the first to be made by an Englishman.[27] Motivating this innovation, Hakluyt states, is its usefulness in the advancement of geographical learning:

> [T]he descriptions of so many parts of the world would farre more easily be conceiued of the Reader [...] that I haue contented myselfe with inserting into the worke on of the best generall mappers of the world onely, until the coming out of a very large and most exact terrestriall Globe [...] composed by. M. Emmerrie Molineux.[28]

In a similar demonstration of the author's awareness of the contemporaneous practical innovations of his subject, the expanded 1600 edition of *Principall Navigations* included a Mercator projection produced by the mathematical geographer Edward Wright.[29] The Hakluytian sensitivity to the new technologies of geographers and cartographers underlines the theory/practice interaction at the heart of 'the descriptions of so many parts of the world'.

   Such an example gives an indication of Bacon's affinities with the prevailing tradition of geographical, scientific writing of the early modern period. In expounding an interdependent link between theory and practice, Bacon sought to challenge what he viewed as the standard ideas of knowledge that had been in existence from the time of Aristotle. While recognising the Greek's eminence as a philosopher and extensively referencing his work, Bacon also aimed to address what he perceived as the ossifying effects of the predominance of Aristotelian modes of learning by the early modern period. In Book One of *The Advancement of Learning,* for instance, Bacon alludes to:

> degenerate learning [that] did chiefly reign amongst the schoolmen, who having sharp and strong wits, and abundance of leisure, and small

variety of reading, but their wits being shut up in the cells of a few authors [...] as their persons were shut up in the cells of monasteries and colleges, and knowing little history, either of nature or time, did out of no great quantity of matter and infinite agitation of wit spin out unto us those laborious webs of learning which are extant in their books.[30]

Such 'schoolmen', suggests Bacon, took 'chiefly Aristotle [for] their dictator'. He rails against cloistered forms of education and Aristotle's prominent role in the 'laborious webs of learning' circumscribing the advancement of science. This rhetoric is advanced further in Bacon's treatise, where the author argues against the stultifying effect of long-established forms of knowledge, specifically in experimental disciplines. Bacon asserts:

And as for the overmuch credit that hath been given unto authors in sciences, in making them dictators, that their words should stand, and not consuls, to give advice; the damage is infinite that sciences have received thereby, as the principal cause that hath kept them low at a stay without growth or advancement. For hence it hath come, that in arts mechanical the first deviser comes shortest, and time addeth and perfecteth; but in sciences the first author goeth furthest, and time leeseth and corrupteth. So we see artillery, sailing, printing, and the like, were grossly managed at the first, and by time accommodated and refined; but contrariwise, the philosophies and sciences of Aristotle, Plato, Democritus, Hippocrates, Euclides, Archimedes, of most vigour at the first, and by time degenerate and imbased: whereof the reason is no other, but that in the former many wits and industries have contributed in one; and in the latter many wits and industries have been spent about the wit of some one, whom many times they have rather depraved than illustrated; for, as water will not ascend higher than the level of the first spring-head from whence it descendeth, so knowledge derived from Aristotle, and exempted from liberty of examination, will not rise again higher than the knowledge of Aristotle.[31]

Recognising the advancement of practical subjects such as 'artillery, sailing, printing, and the like' and how they have come in time to supersede established epistemologies, Bacon employs the expressive aqueous trope of a fountain spring to expound how all types of learning must flow into each other.

As a consequence, in the Baconian model of learning, inter-disciplinarity is promoted, not just across different subjects but also, importantly, different methodologies of investigation. Writing of what he calls the 'three diseases of learning', Bacon reasons: 'Another error [...] is, that after the distribution of particular arts and sciences, men have abandoned universality, or "philosophia prima", which cannot but cease and stop all progression'. To overcome this lack of progression, scholars must assume a more holistic approach to learning to progress their cognitive development and facilitate new 'discoveries'. 'For no perfect discovery can be made upon a flat or a level', Bacon states, 'neither is it possible to discover the more remote and deeper parts of any science if you stand but upon the level of the same science, and ascend not to a higher science'.[32] A primary result of this manifesto is the promotion of multi-disciplinary learning, a key theme of the cosmographical discourses examined in this study. The interweaving by Bacon of theory and exercise and the articulation of science as both meditative and operative would emerge as the touchstones of the Baconian scientific mode. Bacon's writings had especial purchase among scientific thinkers in the seventeenth century, and many stressed two specific fundamentals of Bacon's thought – science as both a theoretical and practical discipline; and the associated emphasis on the vital requirement for instrumentation in empirical inquiry.

The 'Experimental philosopher' and the royal society

The prominence of these features in Baconian science may be further detected in those who characterised themselves as Bacon's most devoted adherents: the founders of The Royal Society. Established in November 1660, The Royal Society of London for Improving Natural Knowledge (to give the organisation its full title) was England's first state-funded scientific institution, and its creators figured Bacon as its spiritual progenitor. As Antonio Pérez-Ramos remarks, the emergence of the Royal Society marks the high watermark of Bacon's popularity as a theoretical scientist: '[T]he apotheosis of Bacon's thought in England, as something approaching a theory of knowledge or a "philosophy of science" coincides with the foundation of the Royal Society and the granting of Royal Charters in 1662-3'.[33] Set up with the intention of promulgating science based upon 'observation and experiment', the Society sought to propagate co-operation and co-ordination in scientific learning.[34] Illustrative of these aims is Henry Oldenburg, one of the earliest secretaries to the Society, who advises in a letter to the Dutch alchemist and physician Johannes Helvetius:

> [I]t is now our business, having already established under royal favour this form of assembly of philosophers who cultivate the world by means of observation and experiment [...] to attract to the same purposes men from all parts of the of the world who are famous for their learning, and to exhort those already engaged upon them to unwearied efforts.[35]

Many figures engaged in the Society's inauguration and initial activities foregrounded their intellectual debt to their predecessor. Bacon's towering presence in the minds of the Society's founders is perhaps most forcefully articulated in an introductory poem by Abraham Cowley to Thomas Sprat's *History of the Royal Society* (1666). Sprat's account, according to Antonio Pérez-Ramos, presents the 'canonical version of Baconianism'.[36] Cowley's prefatory contribution lyricises Bacon in grandiloquent, biblical terms as a forebear of a new type of scientific enquiry:

> Bacon, like Moses, led us forth at last,
> The barren Wilderness he past,
> Did on the very Border stand
> Of the blest promis'd Land,
> And from the Mountains Top of his Exalted Wit,
> Saw it himself, and shew'd us it.[37]

The frontispiece to Sprat's *History* compounds the exaltation of the part played by Bacon in the formation of the institution, depicting him alongside the figures of Charles II, the Society's original patron, and Prince Rupert, the Society's first president. Replicating this pre-eminence among the scientific community, the frontispiece of the Royal Society's self-fashioned authoritative historical narrative places Bacon above the legend 'artium instaurator' or 'instigator of skills'.[38]

One of the main principles of the Royal Society's professed ideal model of science was a stress on the fundamental necessity for instrumentation. Mirroring the Baconian dictum concerning the requisiteness of instruments to investigative inquiry, the covering image to the *History* depicting Bacon incorporates scientific instruments, including globes, quadrants and scales. Furthermore, the account of Bacon's influence in the creation of the Royal Society makes a point of singling out his 'experimental philosophy' and remarking upon his activities as an experimental scientist. Bacon is named as one of a number of thinkers who 'not only disagreed from the *Antients*, but have also propos'd to themselves the right course of slow, and sure Experimenting'. This 'slow, and sure Experimenting' elevates Bacon above his predecessors in Sprat's eyes:

> And of these, I shall onely mention one great Man, who had the true Imagination of the whole extent of this Enterprize, as it is now set on foot; and that is, the *Lord Bacon*. In whose Books there are every where scattered the best arguments, that can be produc'd for the defence of Experimental Philosophy; and the best directions, that are needful to promote it. All which he has already adorn'd with so much Art; that if my desires could have prevail'd with some excellent Friends of mine, who engag'd me to this Work: there should have been no other Preface to the *History* of the *Royal Society*, but some of his Writings.[39]

As this passage attests, a significant trait of Bacon's character for his adherents was his dual status as both deep thinker and practical experimenter. 'The intent of [Sprat's *History*]', summarises Rose-Mary Sargent, 'was to present an account of a unified society dedicated to the disinterested pursuit of knowledge'.[40]

While Sargent declares that such 'polemics' were decidedly dubious – the Society was 'neither as unified nor democratic as Sprat claimed' and the membership was almost exclusively aristocratic – Bacon's perceived character as an experimental philosopher proved a persistent attraction for other Royal Society members such as Robert Boyle, Robert Hooke, William Petty and Christopher Wren.[41] Boyle, in particular, reiterated Sprat's stress on Bacon's empirically inclined science. A Dublin-born natural philosopher whose experiments in chemistry, physics and hydraulics made him one of the foremost scientists of the age, lauded Bacon, carrying out some of the experiments detailed in the *Novum Organum*. In a treatise entitled *The Origin of Forms and Qualities According to the Corpuscular Philosophy*, published in 1666, Boyle lauded Bacon ('Verulam') alongside Lucretius, Basso and Descartes in a list of 'those excellent and especially those modern authors that have professedly opposed the Aristotelian physics'.[42] On the many occasions in which Boyle mentions the work of his predecessor, he frequently adjoins to Bacon's name adulatory adjectives – *Experimental Notes* (1675) makes reference to 'our justly famous Verulam' and his 'short, but excellent, Paper de forma calidi [form of heat]'.[43] Later in his Discourse of Things Above Reason (1681), Boyle cites the 'boldness of our excellent Verulam' in setting forth his 'four sorts of idols [...] those notions that, though radicated in the very nature of mankind, are yet apt to mislead us'.[44]

Boyle's celebration emphasises the widespread perception of Bacon as an overwhelmingly experimental scientist. His description of Bacon as 'one of the first and greatest Experimental Philosophers of our Age' repeats Sprat's designation and draws attention to the status of Bacon in the eyes of the intellectual community as both a philosopher and experimenter whose work stressed the inextricable link between both theory (philosophy) and practice (experimentation) in the carrying out of good science.[45]

'a man seen and expert in cosmography and navigation': Baconian Cabot

Importantly for our consideration of Bacon's engagement with geography, the twin pillars of Baconian science celebrated by men such as Sprat and Boyle – investigation by practical as well as introspective means – may be identified in his discussions of the act of world describing. Moreover, his belief in the benefits of indisciplinarity, that 'each field of theoretical study and its consequences ought to be closely associated' is replicated in his deliberations of geographical science.

In this respect, Bacon conforms to the tradition that saw the collaboration between men of action, artisans and 'mathematical geographers' in geographical science. In Bacon, as in many writers examined previously, the relationship between studious learning and vocational exercise is foregrounded, alerting us to ways in which geographical science served as a perfect example of prevailing epistemological ideas. The theoretical and practical interaction at the heart of good geography is transmitted through descriptions of explorers and maritime discovery; and concomitantly via descriptions of the real application of acquired geographical knowledge.

One of the most conspicuous cases of this technique features in Bacon's *The Historie of the Reigne of King Henry the Seuenth* (1622). Here, the author recounts the adventures of the renowned English explorer Sebastian Cabot, introducing him in terms that yoke his nautical adventuring to an academic erudition: 'a *Venetian,* dwelling in *Bristow,* a man seen an expert in Cosmography *and* Navigation'.[46] A later edition of this history is supplemented with reference to the same figure who, Bacon informs the reader, as part of a 'memorable Enterprize' was awarded an annuity by Edward VI 'for his considerable skill in Cosmographie and the Art of Navigation'.[47] Expert in both cosmography and navigation and intrepid maritime explorer, Bacon's Cabot emblematises the 'dynamic tension' between scholar and artisan outlined by Lesley Cormack.[48] In his portrayal of Cabot – characterised as a commercial sailor who is also patronised by the monarch and is considered an 'expert in Cosmography and Navigation' – Bacon presents a figure who combines 'classical, theoretical, and mathematical roots' with 'economic, nationalistic and practical endeavours'.

Bacon's description of Cabot not only provides an illustration of the interaction outlined by Cormack between 'the world of the scholar [...] and the world of the artisan', but also shows how easily such spheres overlapped in the actions of early modern intellectuals. The writer purposively portrays a figure who combines his mapping knowledge with map production. In the process, Bacon shows how readily the geographically minded employed their learnedness to active processes: upon return from a 'voyage of discoverie' to the New World, Bacon records that Cabot produced a cartographic record of his travels. 'Hee sayled (as hee affirmed at his Returne, and made a Card thereof) very farre Westwards', affirms Bacon, 'with a Quarter of the North, on the North-side *of* Tierra de Labrador, vntill hee came to the Latitude of sixtie seuen Degrees and an halfe, finding the Seas still open'.[49] This account of Cabot's voyage is almost certainly drawn from the third volume of Raphael Holinshed's popular *Chronicles* (1586), which in turn quotes Humphrey Gilbert's *A discourse of a discouerie for a new passage to Cataia* published in 1576, co-authored with George Gascoigne.[50] Leonard F. Dean notes that in writing *The Historie of the Reigne of King Henry the Seuenth*, Bacon was 'forced to rely chiefly on previous literary chronicles, Andrea, Vergil, Hall, Holinshed, Stow, and Speed'.[51] Revealingly, however, neither of the primary source texts for Bacon's account – Holinshed and Gilbert – mentions Cabot's cartographic initiative after his voyage, alluding only to his renown as a map-maker beforehand. Holinshed acknowledges Cabot's geographical learning:

> himselfe to be expert in knowledge of the circuit of the world, and Ilands of the same, as by his charts and other reasonable demonstrations

he shewed [and] caused the king to man and vittell a ship at Bristow, to search for an Iland which he knew to be replenished with rich commodites.[52]

Bacon embellishes his source, making specific reference to his post-voyage cartographic activities, 'as hee affirmed at his Returne, and made a Card thereof'. Cabot, expert in cosmographic knowledge and skilled in maritime exploration as learning in action, stands as an indicative case of Bacon's overarching idea of 'operative science'. His voyage, aided by a theoretical awareness of geographical principles, cosmography and navigation, in their turn manifest a contribution to cartography. Cabot is, in Bacon's account, figured as a proto-Baconian: a scientist in both method and exercise, with each aspect informing and augmenting the other. His voyages act as a physical mode of discovery. However, this discovery is underpinned by a theoretical foreknowledge of the principles of navigation and cosmography; and only fully realised by its subsequent detailing and recording in the form of cartographic production. Cabot concurs with the Baconian dictum that 'no perfect discovery can be made upon a flat or a level, neither is it possible to discover the more remote and deeper parts of any science if you stand but upon the level of the same science'.[53] In his endeavours, tellingly augmented by Bacon from the source texts, Cabot is portrayed as utilising principles and action to grasp a greater comprehension of the 'parts of science' or knowledge on which he stood. Geography, emblematised in its practitioner Cabot, embodies a significant facet of the Baconian scientific mode.

Instruments in the geography of Salomon's house

If the portrayal of the geographer and voyager Cabot personifies a singular model of the Baconian 'operative science' incorporating theoretical and applied learning, the writer's brief discussions of geography also encompassed another vital aspect of his thought – the requirement of instrumentation in scientific enquiry. The articulation within geographical texts of the need for instrumentation was a recurrent theme throughout educational treatises in the early modern period. Importantly, such a stress recurs in Bacon's reflections on an idealised manner of experimentation and discovery. The necessitation of instrumentation in cartography and its ancillary disciplines was cited as prototypical of science as a whole. While Cabot served as a proto-Baconian in his geographical activities, geography itself was characterised as a model for Baconian scientific methodology as a whole.

Instruments are fundamental to Bacon's ideal scientific mode. Alongside Sprat's portrayal in his *History* of Bacon in a laboratory teeming with experimental instruments, more recent scholars of Bacon's work have stressed the writer's emphasis on properly equipped and standardised scientific workshops. Paolo Rossi, for instance, in emphasising the twin impulses of 'institutionalisation and professionalization' in modern science, has maintained that Bacon's ideas about natural philosophy and how it should be studied remain formative. 'Bacon', observes Rossi:

> is one of the constructors – perhaps the greatest – of that which can be called a modern image of science. His discourse on this theme is ample, articulated, full of intellectual force, literarily efficacious, rich in imitable metaphors. His discourse does not concern only the method of science (everyone knows that he made an important contribution to the discussion on induction). It concerns above all the function of science in human life, the ends and values that must characterize scientific knowledge; it concerns that which today we would call an ethics of research; it concerns, finally, the ways in which this form of knowledge must present itself in comparison to the other forms of cultural life: poetry, history, religion, ethics, politics.[54]

The drive towards what Rossi describes as 'professionalisation' and 'institutionalisation' in Baconian logic is most palpable in the brief manifesto of Salomon's House, the scientific institution in the Utopian travel narrative *The New Atlantis* (1627). The body responsible for the functioning of Salomon's House is motivated by a clear, pre-established set of aims: 'The End of our Foundation', explains the guide to Bacon's narrator, 'the knowledge of all causes, and the secret motions of things', and the 'enlargening of the bounds of Human Empire'. Conveying the institutional character of the research project, the processes and systems accompanying its endeavours further establish the institute's professionalisation – geared towards a single profit ('the affecting of all things possible') Salomon's House contains a rigidly systematised hierarchy of scientific vocations.[55]

The importance of instruments and their utility to Bacon's notions of scientific enquiry is outlined in the stated rubric of Salomon's House. The methodology sustaining the attempt to 'learn the knowledge of all causes, and the secret motions of things' for the 'enlargening of the bounds of Human Empire' is related by Bacon's guide and commences with the revealing double formation '[t]he preparation and instruments are these'.[56] From the outset, instruments are figured as indispensable elements of scientific inquiry. Consequently, a multiplicity of instruments fill the many wondrous towers, caves, rooms and laboratories of the House as the building pullulates with object-laden experimenters. There are 'engines for multiplying and enforcing of winds', 'artificial wells and fountains', 'divers mechanical arts', 'instruments which generate heat by motion', microscopes or 'glasses and means to see small and minute bodies perfectly and distinctly', 'divers instruments of music', 'engines and instruments for all sorts of motions' and 'instruments of war and engines of all kinds'.[57]

The prevalence of instrumentation in Salomon's House is not expressly novel in the context of Bacon's overall view of the scientific venture.

Nonetheless, the list of 'instruments' and 'engines' (accompanied by an enumeration of the individual 'houses' in the institution) is symptomatic of Bacon's long-held convictions concerning how science and the discovery of 'the knowledge of Causes' should be carried out. Bacon's explicit inclusion of instrumentation in each room in the House and its implicit emphasis on the fundamental role of instruments in science in his description is not entirely novel.[58] It reiterates the themes of earlier treatises, not only in his 'Councillor Advising the Study of Philosophy' but also his letter to Henry Savile, which stressed scholarly study, experiential learning and familiarisation with the experiments and instruments of selected disciplines (as in use of the archery analogy which alludes to a 'stronger bow').

Susan Bruce, being careful to address the complex intertwining themes of secrecy, science and power in the Bensalemite society surrounding the House and the ambivalent status of the institution as an ameliorating force in 'the lives of "men", of ordinary people', asserts that 'contemporary readers of the *New Atlantis* [...] saw [Salomon's House's] purpose as being that of providing a blueprint for a new scientific institution'.[59] However, Salomon's House serves as a striking symbolic representation of the Baconian stress on the importance of instrumentation to a multitude of scientific disciplines.

Advancing learning through geography

In addition to featuring in the characterisation of Cabot and Salomon's House, the Baconian formulation of geography as a paradigmatic scientific subject features in *The Advancement of Learning*. In this work, the requirement of the instruments of mapping to geographical learning as articulated in geographical texts served as an exemplar for Bacon in his deliberations on a paradigmatic scientific method. In a discussion about the study of geography near the start of book two, Bacon states:

> But certain it is, that unto the deep, fruitful, and operative study of many sciences, specially natural philosophy and physic, books be not only the instrumentals; wherein also the beneficence of men hath not beene altogether wanting: for we see, spheres, globes, astrolabes, maps, and the like, have been provided, as appurtenances to astronomy & cosmography, as well as books: we see likewise, that some places instituted for physic, have annexed the commodity of gardens for simples of all sorts, and do likewise command the use of dead bodies for anatomies.[60]

Bacon assigns to the 'study of many sciences' three discrete descriptives: 'deepe', 'fruitfull' and 'operatiue'. When *The Advancement of Learning* first appeared (1605), the phrase 'operative' was commonly suggestive of something useful or active. In an English translation of the French humanist scholar Louis Le Roy's version of Aristotle's discourses of government, published in 1598, a discussion of the structure of the family and the role of servants within the household refers to 'operative and working instruments',[61] while in a later translation of Plutarch by Philemon Holland (1603) the author writes:

> The doctrine of Philosophy is not like unto an imager who casteth dumbe and deafe idole statues, without any sense, onely for to stand upon a base as *Pindarus* was woont to say, but is willing to make whatsoever it toucheth, active, operative and lively.[62]

The parallel Bacon draws between 'the deep, fruitful, and operative study of many sciences' and 'Astronomy and Cosmography' shows his awareness of the subject's focus on the practical as well as the philosophical. Bacon repeats the familiar comparisons between geographical science and anatomy – as noted, mapping and anatomisation were frequently associated in their vocabulary and methodologies by early modern thinkers. Bacon taps into this discourse which was frequently marked by a promotion of interdisciplinary approaches and derived from a deep appreciation of the benefits of practical endeavour. William Harvey, influential scientist and one-time doctor to Bacon, exemplifies the strident practicability of the seventeenth-century anatomist: 'I profess both to learn and to teach anatomy', he confirms in the preface to *De Motu Cordis* or *The Motion of the Heart and Blood in Animals* (1628), 'not from books but from dissections'.[63]

The citation in *The Advancement of Learning* of geography's consanguinity to anatomy further indicates Bacon's focus on the necessity for instrumentation in geographic science. '[S]pheres, globes, astrolabes, maps, and the like' may be categorised by Bacon as 'appurtenances', but they are appurtenances whose importance equals that of 'bookes' in the pursuit of learning. The importance of cartographic instrumentation for the study of geography in the Baconian model is indicated by a comparison with the vital worth of corpses in anatomical procedure: just as learning about bodies requires the use of cadavers, so cartographic instruments are fundamental to the pursuit of geographical knowledge.

Thus, the two precepts central to Bacon's understanding of geography – firstly, the symbiosis of theoretical learning and practical implementation and the fundamental importance of instrumentation to the subject is crystallised by Bacon in the next few sentences of book two of *The Advancement of Learning*:

> In general, there will hardly be any main proficience in the disclosing of nature, except there be some allowance for expenses about experiments; whether they be experiments appertaining to *Vulcanus* or *Daedalus,* furnace or engine, or any other kind; and therefore as secretaries, and spials of Princes and States bring in Bills for Intelligence; so you must allow the spials and intelligencers of nature, to bring in their bills, or else you shall be ill advertised.[64]

Bacon deals with one of the foremost issues of the formation of a scientific 'institute' or dedicated place of learning – expenditure. 'The disclosing of nature' and studies 'appertaining to *Vulcanus* or *Dedalus,* Furnace or Engyne, or any other kind', is underpinned by 'experiments' and heavily dependent upon the presence of scientific equipment in the sphere of discovery. As with the geography/anatomy parallel, Bacon argues by comparison: fiscal expenditure on experiments is as vital to science as 'Bills for intelligence' are to the functioning and administration of the state. If Cabot was a proto-Baconian scientist; and the 'mathematical house' with its geographically inflected activities evoked the necessitous application of both theory and practice with instruments of Baconian science; then geography, more generally, with its obligation for instrumentation, a combination of learning and experimentation was a proto-Baconian science. Paralleling the prevailing belief in geography as a theoretical and practical discipline, Bacon's wider reflections upon geography and its implementation are characteristic of the writer's broader programme for experimental science. Mapping in the period figured not just as a medium of social, political and colonial expression – its very processes functioned as a prototype for learning itself.

Notes

1. *Troilus and Cressida* (1.3.205).
2. Francis Bacon, *The Major Works*, ed. Brian Vickers (Oxford: Oxford University Press, 2002), p. xx. According to Guido Guglioni: 'In traditional accounts of early modern philosophy, Bacon's contribution has often been confined to the field of epistemology and scientific methodology but his engagement with philosophical ideas was in fact wider and richer' Guido Giglioni, 'Francis Bacon', *The Oxford Handbook of British Philosophy in the Seventeenth Century*, ed. Peter R. Anstey (Oxford: Oxford University Press, 2013), p. 49.
3. Antonia McLean, *Humanism and the Rise of Science in Tudor England* (London: Heinemann, 1972), p. 232.
4. According to Markku Peltonen, the 'traditional story' of Bacon's death relates that 'in an unseasonably cold spring it had occurred to Bacon to test whether snow would preserve flesh from putrefaction, as salt does. In order to conduct the experiment, he obtained a hen and stuffed it with snow with his own hands, but also caught the chill from which he subsequently died'. Markku Peltonen, 'Bacon, Francis, Viscount St Alban (1561–1626)', *ODNB* (Oxford: Oxford University Press, 2004), online edn. January 2008, http://www.oxforddnb.com/view/article/990 (accessed 11th April 2021). This legend – with its romanticised inference that Bacon, the ultimate empirical scientist, died in pursuit of scientific discovery – has since been challenged. See Lisa Jardine and Alan Stewart, *Hostage to Fortune: The Troubled Life of Francis Bacon, 1561–1626* (London: Gollancz, 1998) pp. 502–508. John Dee, *The elements of geometrie of the most auncient philosopher Euclide of Megara* (London: John Day, 1570), sig A4r.
5. As Jonathan M. Smith points out, 'in his essay on travel, Francis Bacon made attention to the geographical features of foreign realms a form of public service'. Smith, 'State Formation, Geography, and a Gentleman's Education', *Geographical Review*, 86:1 (1996), p. 95.
6. Bacon, *The Essays* (London: Penguin, 1985), p. 148.

7. See, for example, 'Of Empire' and 'Of Plantations', *The Essays*, pp. 115–119, 162–164; *The Charge touching duels* (1614), *The Major Works*, pp. 304–313; and *An Advertisement touching the Controversies of the Church of England* (c.1589–1591), *The Major Works*, pp. 1–19.

8. An indicator of the extent of Bacon's writings is the current British Academy Oxford Francis Bacon Project, which aims at replacing 1900 *The Works of Francis Bacon by Francis Bacon*, edited by James Spedding, Robert Leslie Ellis and Douglas Denon Heath, with a new 15-volume scholarly edition. See https://onlinebooks.library.upenn.edu/webbin/metabook?id=worksfbacon (accessed 23rd November 2020).

9. Bacon, *Of the Advancement and Proficience of Learning, The Major Works*, p. 151.

10. Bacon, *New Atlantis, The Major Works*, p. 486.

11. For a discussion of the extent of Bacon's classification of scientific and philosophical epistemology, see Sachiko Kusukawa, 'Bacon's Classification of Knowledge', *Cambridge Companion to Francis Bacon*, ed. Markku Peltonen (Cambridge: Cambridge University Press, 1996), pp. 47–74.

12. Alvin Snider, 'Bacon, Legitimation and the "Origin of Restoration Science"', *The Eighteenth Century*, 32:2 (1991), p. 119. For an elaboration of Voltaire's view that Bacon was 'a great philosopher, a good historian, and an elegant writer', see Voltaire, 'On Chancellor Bacon', *Philosophical Letters: Letters Concerning the English Nation*, trans. Ernest Dilworth (New York: Dover, 2003), pp. 46–51. For Theodor Adorno and Max Horkheimer's analysis of Bacon's ideas, see Adorno and Horkheimer, 'The Concept of Enlightenment', *Dialectic Enlightenment: Philosophical Fragments*, ed. Gunzelin Schmid Noerr, trans. Edmund Jephcott (California: Stanford University Press, 2002), pp. 1–34, especially pp. 1–5.

13. William Rawley, *Resuscitatio, or, Bringing into publick light severall pieces of the works, civil, historical, philosophical, & theological, hitherto sleeping, of the Right Honourable Francis Bacon, Baron of Verulam, Viscount Saint Alban* (London: Sarah Griffin, 1657), sig. B2r. Rawley erroneously dates his subject's birthday – Bacon was born in 1561. Peltonen, *ODNB*.

14. Markku Peltonen, 'Introduction', *Cambridge Companion to Bacon*, ed. Markku Peltonen (Cambridge: Cambridge University Press, 1996), p. 1. For a flavour of the discussions of Bacon's contribution to science, or perceived lack thereof, see Lisa Jardine, *Francis Bacon: Discovery and the Art of Discourse* (Cambridge: Cambridge University Press, 1974) and Antonio Pérez-Ramos, *Francis Bacon's Idea of Science and the Maker's Knowledge Tradition* (Oxford: Clarendon Press, 1991).

15. Pérez-Ramos, *Francis Bacon's Idea of Science*, p. 7.

16. Pérez-Ramos, *Francis Bacon's Idea of Science*, p. 7.

17. Markku Peltonen, 'Bacon, Francis, Viscount St Alban (1561–1626)', *ODNB* http://www.oxforddnb.com/view/article/990 (accessed 14th July 2020).

18. Bacon, 'Of Studies', *The Major Works*, p. 439.

19. Bacon's letter to Savile, despite the early glimpse it offers into Bacon's developing ideas surrounding knowledge and learning, is relatively absent from the scholarly discussion, being cited only once in the collection of essays comprising *The Cambridge Companion to Bacon*, and ignored in other substantial studies of Baconian thought such as Pérez-Ramos, *Francis Bacon's Idea of Science and the Maker's Knowledge Tradition* and Jerry Weinberger, *Science, Faith, and Politics: Francis Bacon and the Utopian Roots of the Modern Age* (Ithaca: Cornell University Press, 1988).

20. Bacon, 'A Letter and Discourse to Sir Henry Savile touching Helps for the Intellectual Powers', *The Major Works*, p. 114. Vickers glosses 'conceit' as 'the faculty of forming ideas, concepts' (p. 573).

21. Bacon, 'A Letter and Discourse to Sir Henry Savile', *The Major Works*, p. 117.
22. Bacon, 'A Letter and Discourse to Sir Henry Savile', *The Major Works*, pp. 118–119.
23. Bacon, *The Major Works*, p. 573.
24. Bacon, 'A Letter and Discourse to Sir Henry Savile', *The Major Works*, p. 114.
25. Jardine, *Francis Bacon*, p. 99.
26. Lesley B. Cormack, Cormack, *Charting an Empire*, p. 137.
27. For an account of Molyneux's globes, see Helen Wallis, '"Opera Mundi": Emery Molyneux, Jodocus Hondius and the First English Globes', *Theatrum Orbis Librorum: Liber Amicorum presented to Nico Israel on the Occasion of His Seventieth Birthday*, ed. Ton Croiset van Uchelen et al. (Utrecht: Hes & De Graff Publishing, 1989), pp. 94–104.
28. Richard Hakluyt, *The principal nauigations, voyages, traffiques and discoueries of the English nation* (London: George Bishop, 1589), sig. *4v.
29. For a discussion about this map and Hakluyt's use of maps more generally, see R. A. Skelton 'Hakluyt's Maps', *The Hakluyt Handbook: Volume 1*, ed. D.B. Quinn (London: Hakluyt Society, 1974), pp. 48–73.
30. Bacon, *The Advancement of Learning*, *The Major Works*, p. 140.
31. Bacon, *The Advancement of Learning*, *The Major Works*, p. 140.
32. Bacon, *The Advancement of Learning*, *The Major Works*, p. 146.
33. Pérez-Ramos, *Francis Bacon's Idea of Science*, p. 4.
34. Marie B. Hall, 'Science in the Early Royal Society', *The Emergence of Science in Western Europe*, ed. M. Crosland (London: Macmillan, 1975), p. 57.
35. Henry Oldenburg quoted in Hall, 'Science in the Early Royal Society', p. 57.
36. Pérez-Ramos, *Francis Bacon's Idea of Science*, p. 99.
37. Cowley in Thomas Sprat, *The History of the Royal Society of London* (London: 1667), sig. B2r. For a further examination of Sprat's *History* and its stylistic and structural parallels with Bacon's work, see H. Fisch and H.W. Jones, 'Bacon's Influence on Sprat's *History of the Royal Society*', *Modern Language Quarterly*, 12 (1951), pp. 399–406.
38. Sprat, *The History of the Royal Society of London*, title page.
39. Thomas Sprat, *The History of the Royal Society of London*, pp. 35–36.
40. Rose-Mary Sargent, 'Bacon as an Advocate for Cooperative Scientific Research', *Cambridge Companion to Bacon*, p. 167.
41. Sargent, 'Bacon as an Advocate for Cooperative Scientific Research', p. 167.
42. Robert Boyle, *Selected Philosophical Papers of Robert Boyle*, ed. M.A. Stewart (Manchester: Manchester University Press, 1979), p. 10.
43. Boyle, *Selected Philosophical Papers*, p. 14.
44. Boyle, *Selected Philosophical Papers*, p. 229.
45. Robert Boyle, *The Christian virtuoso shewing that by being addicted to experimental philosophy* (London: Edward Jones, 1690), p. 8.
46. Francis Bacon, *The historie of the reigne of King Henry the Seuenth* (London: Printed by W.G. for R. Scot, T. Basset, J. Wright, R. Chiswell, and J. Edwyn, 1676), p. 187.
47. Francis Bacon, *The History of the Reigns of Henry the Seventh, Henry the Eighth, Edward the Sixth, and Queen Mary the First* (London: Printed by W.G. for R. Scot, T. Basset, J. Wright, R. Chiswell, and J. Edwyn, 1676), p. 151.
48. Cormack, *Charting an Empire*, p. 26.
49. Bacon, *The History of the Reigns of King Henry the Seventh*, pp. 188–189.
50. Raphael Holinshed, *The Third Volume of Chronicles, Beginning at Duke William the Norman, Commonlie called the Conqueror* (London: Henry Denham, 1586), p. 785.
51. Leonard F. Dean, 'Sir Francis Bacon's Theory of Civil History-Writing', *ELH*, 8:3 (1941), p. 170.

52. Holinshed, *The Third Volume of Chronicles*, p. 785.

53. Bacon, *The Advancement of Learning, The Major Works*, p. 146.

54. Paolo Rossi, 'Bacon's Idea of Science', *Cambridge Companion to Bacon*, pp. 25–26.

55. Bacon, *The New Atlantis, The Major Works*, p. 480.

56. Bacon, *The New Atlantis, The Major Works*, p. 480.

57. Bacon, *The New Atlantis, The Major Works*, pp. 481–486.

58. See also Julian Martin, *Francis Bacon, the State, and the Reform of Natural Philosophy* (Cambridge: Cambridge University Press, 1992), p. 216.

59. *Three Early Modern Utopias: Thomas More's 'Utopia', Francis Bacon's 'New Atlantis' and Henry Neville's 'The Isle of Pines'*, ed. Susan Bruce (Oxford: Oxford University Press, 1999), pp. xxxi–xxxv.

60. Bacon, *The Advancement of Learning, The Major Works*, p. 172.

61. Aristotle, *Aristotles politiques, or Discourses of gouernment*, trans. Louis Le Roys (London: Adam Islip, 1598), p. 22.

62. Plutarch, *The philosophie, commonlie called, the morals vvritten by the learned philosopher Plutarch of Chaeronea*, trans. Philemon Holland (London: Arnold Hatfield, 1603), p. 289.

63. William Harvey quoted in Lisa Jardine, *Ingenious Pursuits: Building the Scientific Revolution* (Norfolk: Doubleday, 1999), pp. 111–112.

64. Bacon, *The Advancement of Learning, The Major Works*, p. 172.

Part III
Maps on stage and page

6   Plotting Marlovian geographies

In a period characterised by an increasing interest in the diversity of geographical disciplines, Christopher Marlowe is perhaps unparalleled in early modern English drama for his engagement with the practice of 'world-describing'. Marlowe's drama persistently theatricalises travel, exoticism and cartography's varieties. The structure of the following interrogation of the multiplicities of Marlovian geography is two-fold. The first section sketches Marlowe's personal engagement with geography through his background, education and university career. In doing so, I contextualise Marlowe's geographical learning by examining his exposure to the ideas, themes and precepts of the art of world describing during his early life. In the second, I explore how Marlowe employs some of the prevalent ideas surrounding geography, for example, the notion of vicarious travel through the activity of map-reading as a dramatic centrepiece to Tamburlaine; and also the paradigm of cosmographical exploration, with its attempts at divine understanding and symbiosis of theory and practical experience, for narrative ends in Faustus. I will argue that while manifesting itself in an encyclopaedic geographical list of names, places and allusions, the scope of Marlovian geography extends beyond superficial enumeration. In his plays, Marlowe engages fully with a number of the core themes prevalent in the geographical imagination of the age, including staging the vicarious journey by map and testing out educational ideas surrounding cosmographical learning.

Marlowe's cartographic gaze

'Our writers and great men had something in them that savoured of the soil from which they grew', insists William Hazlitt of the milieu that dominated English drama in the early modern period, '[t]hey were truly English. [...] The mind of their country was great in them, and it prevailed'.[1] In the case of one of its most fascinating exponents, Marlowe, Hazlitt is wrong – in an epoch characterised by an increasing interest in geography, Christopher Marlowe is perhaps unparalleled in early modern English drama for his engagement with the practice of world-describing and its relentlessly diverse panoramas. If his plays are anything to go by, the playwright appears

DOI: 10.4324/9781003200376-10

largely uninterested in 'the soil from which [he] grew'. Far from 'prevail-ing', England as a geographical entity is often marginalised in Marlowe's plays. Often, that is, where it is not elided completely. The most identifi-ably Marlovian plays – those which exhibit a stress on his famed 'mighty line' and foreground the agency of the central character – were resolutely un-English. In the geography of *Tamburlaine the Great*, *The Jew of Malta* and *Doctor Faustus*, Marlowe opts to disregard homogeneity in favour of variety. The panoramic span of Marlowe's drama is especially evident in their settings: where *Tamburlaine* parts 1 and 2 begin in Persepolis and on the banks of the Danube, respectively, *Faustus* opens in a study in Wittenberg ('Wertenberg'), Germany, *Dido, Queen of Carthage* in North Africa, and *The Jew of Malta* in a Maltese treasury. Even in *Edward II*, an outwardly English drama about an English monarch set largely in England, Marlowe's inclination to geographical breadth is subtly revealed: as well as commenc-ing in France, there is a marginal yet crucial role played by Ireland in the functioning of the play.[2]

Marlowe's oeuvre is therefore ripe for critical consideration of how early modern English culture engaged with the meridians and parallels of geo-graphical science. The structure of this exploration of the multiplicities of Marlovian geography is two-fold. The first section sketches Marlowe's personal engagement with geography through his background, education and university career. In doing so, it contextualise Marlowe's geographical learning by examining his exposure to the ideas, themes and precepts of the art of world describing during his early life. In the second, it explores how Marlowe employs some of the prevalent ideas surrounding geography set out in previous chapters, for example, the notion of mediated travel through the activity of map-reading as a dramatic centrepiece to *Tamburlaine*; and also the paradigm of cosmographical exploration, with its attempts at divine understanding and symbiosis of theory and practical experience, for narrative ends in *Faustus*. I will argue that while manifesting itself in an encyclopaedic geographical list of names, places and allusions, the scope of Marlovian geography extends beyond superficial enumeration. In his plays, Marlowe engages fully with a number of the core themes prevalent in the geographical imagination of the age, including staging the vicarious journey by map and testing out the fertile educational ideas surrounding cosmographical learning.

As Ben Jonson's remark on Marlowe's 'scenicall strutting' reveals, the shifting geography of the Marlovian corpus has been one of the most con-sistently recognisable traits of the playwright's work.[3] The span of Marlowe's locational panoply – reciprocated in his extraordinarily diverse dramatis personae–evidences what Michael Neill has called 'the intoxicated exoti-cism of Marlovian cosmography'.[4] The stress on the exotic in Marlovian drama is significant for our understanding of Marlowe's geographical sen-sibilities as it aligns the playwright with the popular enthusiasms of the age. In late sixteenth-century England, the notion of travel to foreign countries and climes captured the imagination of many. 'Al studies have theyr special tymes', observes Richard Willes in his preface to Richard Eden's popular *History of Travayl* (1577), 'of late who taketh not upon him to the discourse of the whole worlde, and eche province thereof particularly?'[5] Marlowe's plays, with their multi-cultural characters and multiplicity of settings, stage dramatically this discourse.

Marlowe's interest in geographical discourses is especially palpable in the constant travelling narratives of his plays and their wide topographical scope. This range has been seen as a distinctly Marlovian trait, such extensiveness emphasised by its contrast with the relative brevity of his authorial career: 'Marlowe's career as a dramatist was very brief', writes Irving Ribner, 'but it was also very varied'.[6] In one sense, Marlowe's dramatic output reflects the equivalencies between theatrical art and geographical science in the period, most obviously in language. Jonathan M. Smith, surveying the lie of the land in seventeenth-century English geographical education, shows how the practice of geography had an affinity with the stage in the Renaissance:

> The geography of position confers an extraordinary ability to divide and traverse abstracted space; the geography of possession provides an instrument for conceptualising the content of [...] space in terms which are familiar, uniform, and global. Together they provide a systematic model of the earth as an intelligible object of inquiry and a practicable theatre of activity.[7]

What Smith terms 'the geography of position' may be understood more specifically as the surveying, drawing and dissemination of cartography. The linguistic coincidences flagged up by Smith between such a process and the artistic endeavour of Marlowe – the 'theatre of activity' – are discernible, most obviously in the various theatres of both cartographer (Speed and Ortelius, for example) and playwright (the Globe theatre which originally staged many of Shakespeare's plays). Smith's description of the coeval geographies of 'position' and 'possession' and his allusion to 'the ability to divide and traverse abstract space' through cartography points us towards the similarities between geographical science and dramaturgy. For example, *Tamburlaine*, which was performed by the Lord Admiral's Men at the Rose Theatre on Bankside in 1587, presents a globe-trotting adventure across three separate continents, traversing the stage of the world in a similar fashion to the way in which the map-reader might travel through the map of the world.[8]

Marlowe's medium, then, shared affinities with cartography: playwrights, like mapmakers, littered their scenes with names and exotic places. The penchant for foreign locales in English theatre was well established by the late sixteenth century. Earlier popular works such as *Fedele and Fortunio* (1583) and John Lyly's *Sappho and Phao* (1584) were sited in non-English settings (Naples and Sicily, respectively), presenting to early modern English theatre-goers foreign characters with unusual names and alien customs. Thomas Kyd's *The Spanish Tragedy*, approximately contemporaneous with Marlowe's play, is exemplary of this particular dramatic typology: concerning a peregrinatory tale of love, death and sex set primarily in Spain, it also encompasses Europe, Asia and the New World.

However, even within such a tradition, Marlowe's plays retain a distinctiveness in the diversity of their ethnic, cultural and geographical diversity. Revealingly, Marlovian drama, and *Tamburlaine* in particular, has been identified as one of the most influential prototypes for what David McInnis has described as the emerging genre of 'voyage drama'. 'Marlowe's *Tamburlaine* plays [...], if not the first exotic voyage dramas', McInnis observes, 'were certainly the most influential and successful of the early exponents of the form'.[9] McInnis helpfully defines the voyage drama genre as 'plays which incorporate scenes of travel, deploy genuinely exotic settings which are not mere foils for London, or are in some way concerned with the motivation and the consequences of travel'.[10] The 'motivation' to travel is embedded in the plot of the play: like the 'vagrant ensign' of its title, the action is marked by a ceaseless kinetic energy.[11]

Moving across the stage and the map

Tamburlaine's vigour is representative of the broader theme of movement in Marlowe's work and the central importance of geography in the operation of his theatricality: *Faustus* circles Europe in a presage of the Grand Tour, *Dido, Queen of Carthage* the Mediterranean and *Edward II* various topographies around the Atlantic archipelago, from Wales to Gloucestershire and Warwickshire. 'At once it seems [Marlowe] delights in his heroes' movement', writes Mark Thornton Burnett, 'in Barabas's description of his Machiavellian practices in Italy, France and Germany, or Faustus's travels with Mephistopheles to far corners of the globe'.[12] Corollary to this delight, the necessity for diverse locations underlines Marlowe's geographical range. Much like William Cuningham, who celebrated the 'cosmographicall glasse' wherein 'we may beholde the diuersitie of countries: natures of people, and innumerable formes of Beastes, Foules, Fishes, Trees, Frutes, Stremes, & Meatalles', the playwright revels in the linguistic richness and exoticism of a varied geographical purview.[13] This is apparent in the play's citation of a vast array of toponyms. Emrys Jones counts upwards of 120 different place names in the text of both the original *Tamburlaine* and its sequel, a figure easily exceeding similar plays in the period.[14]

While Marlowe's variety in character and setting is striking, scholars have also been alerted to the level of geographical learning evident in Marlowe's plays. Ethel Seaton, in essays such as 'Marlowe's Map' (1924) and 'Fresh Sources for Marlowe' (1929) has shown the extent of Marlowe's awareness of contemporary maps and atlases.[15] Emrys Jones, meanwhile, has accentuated the importance of Seaton's cartographically centred work for scholars of Marlowe and, as a consequence, Marlowe's map-mindedness for the wider understanding of his work:

> [Seaton has] demonstrated conclusively that one notable passage in Part Two was not a mere fantasy of outlandish names but an account of a journey (by Techelles) through Africa, which Marlowe had based on a close study of Ortelius's *Theatrum Orbis Terrarum*. This great world atlas was first published in Antwerp in 1570. Some of Marlowe's details, which had been taken by editors to be errors or blunders on his part, could now be justified by reference to Ortelius, notably the placing of 'Zanzibar' on the west coast of Africa rather than on the east. But the chief effect of Seaton's work was to make Marlowe's procedure in the Tamburlaine plays seem incomparably more rational than before. It showed him to be aware of some of the most recent technological innovations by putting him in touch with the enormously influential map culture of his time.[16]

Marlowe's variegated geographies are entrenched in the identity of his main protagonists and represent a central aspect of the dramatic structure of his plays. One of his most sensitive readers, Stephen Greenblatt, has explicitly linked Marlowe's geographical scope with the playwright's formulation of characters. Drawing in his own notion of Marlovian 'absolute play', Greenblatt writes that '[i]dentity is a theatrical invention that must be reiterated if it is to endure'.[17] In plays such as *Tamburlaine* and *Edward II* characterisation co-exists symbiotically with the heterogeneity of Marlovian geography. The shifting landscapes in *Tamburlaine* accompany the dynamism of the main character. His action and his identity as a dramatic character, as Greenblatt rightly notes, are partly expressed through movement: 'Tamburlaine no sooner annihilates one army than he sets out to annihilate another'.[18] Geographical variety lies at the centre of Marlowe's dramatising of a 'struggle against extinction'. As a consequence of this attempt to avoid stasis, there is in Marlowe's drama a persistent propulsion from exotic location to exotic location.

Schizophrenia and displacement: Marlowe's geography

Situated at a point of knowledge transformation, or more precisely, slippage from the traditional into the novel in geographic discourse, Marlowe is important for our consideration of early modern English drama's engagement with mapping because he is witness to a crucial point in the narrative of geographical and cartographic history. As readers have noted, his drama evidences a writer working simultaneously in the death throes of an 'old' geography and in the birth of a 'new' geography and receptive to the attendant intricacies, precepts and themes of both. John Gillies, for example, identifies an almost destabilising sense of geography in Marlowe's dramaturgy which reflects the epistemological discombobulations of the age. *Tamburlaine,* for example 'manifests – with a power unsurpassed by any other Renaissance geographical or "poetic geographic" text – the schizophrenia of the Renaissance geographical imagination caught (as I see it) between the amoralism of the New Geography, and the moralism of the old'.[19] Garrett Sullivan, in a broader survey of the Marlovian canon, has concurred with Gillies' assessment. Sullivan asserts that Marlowe's engagement with geographical discourse and its ideas was conducted within an 'epochal moment in the histories of geography and cartography – that of the emergence of the "new geography"'. Sullivan continues:

> This moment is understood as marking the turning point from an imprecise and religious or mythopoetic geography to an accurate and scientific one – from, for example, the medieval map centred on the sacred site of Jerusalem to the famous cartographic projection associated with the atlas-maker Gerard Mercator, which allows for the representation of space as homogeneous and uniformly divisible. Characterised by the proliferation of increasingly precise representations of the world (with Ortelius's atlas being a prime example), the new geography was made possible by a number of historical phenomena, such as improved mapping technologies; the growing desire and need for accurate cartographic information; and the ever-widening distribution of printed geographical materials including maps and atlases.[20]

This view alerts us to Marlowe as the dramatic geographer at a liminal point in the history of geography, revelling in the contemporary late-sixteenth-century imbrication of the emergent practical geographical enterprise ('spheres, globes, astrolabes, maps and the like'), and the 'mythopoetic geography' of more established ideas of geographical representation (the overtly religious T-O maps of medieval cartography). Where Gillies suggests that Marlowe encapsulates the 'schizophrenia of the Renaissance geographical imagination' and imbues his work with a tension between 'old' and 'new' moralities, Sullivan states much more categorically that '[i]t is of the new geography that Marlowe's plays appear to be such a conspicuous product'.[21] Elsewhere, Lisa Hopkins links Marlowe's geography more specifically with 'questions of religious belief'.[22] According to Hopkins, the undermining of pre-conceived religious ideas by geography in the Renaissance – Hopkins specifically cites the discovery of America as 'precipitat[ing] the great crisis of faith which ultimately produced the Reformation, since the failure of the Bible to mention the New World cast doubt on the supposed omniscience of the Scriptures'[23] impressed itself deeply on Marlowe, engendering a sincere belief in both 'spiritual' and 'physical' geographies. 'For Marlowe', Hopkins writes:

> Knowledge of geography gives access to the contours of the next world as well as the present one – and as the present one expands, the imaginative space allotted to the next one visibly shrinks and withers.[24]

Identification of these various geographies – 'old', 'new', 'schizophrenic', 'spiritual', 'imaginative' – amply illustrates the richness of Marlowe's geographical sensibility.[25]

'The fruitful plot of scholarism': Marlowe's geographical education

The roots of these qualities in Marlovian geography can be discerned in the education of the playwright. Geographical diversity was present in Marlowe's life from an early age. Canterbury, Marlowe's birthplace, was one of the foremost sites of pilgrimage from the medieval period onwards, drawing visitors from across the British Isles and the Continent – throughout the Middle Ages, Canterbury retained an attraction for pilgrims rivalled only by Rome, Jerusalem and Santiago.[26] Even after the Reformation and its transformative effect on religious pilgrimages in England, the populace of Marlowe's home town maintained a degree of ethnic and religious heterogeneity.[27]

Awareness of ethnological geographical multiplicity was supplemented by an education rich in geographical knowledge. By the late sixteenth century, influential pedagogical tracts, as we have seen, propagated the notion of geographical science as part of the study of 'cosmography', a kaleidoscopic subject which encompassed a vast and often contradictory rubric. This stemmed from continental educationalists such as Desiderius Erasmus, Juan Luis Vives and Leon Battista Alberti.[28] For example, in *Della Famiglia* (1434), a tract described by Kenneth Charlton as Alberti's 'championing of the personal, and social ideals of civic humanism', the Italian writer summarises the paradigmatic education of the humanist scholar, notably emphasising, among other disciplines geography.

In approaching Marlowe's educational experience of geography, the writings of home-grown theorists such as Thomas Blundeville are instructive. According to Blundeville's popular textbook, *His Exercises* (1594):

> [Cosmography is] the description of the whole world, that is to say, of heauen and earth, and all that is contained therein. What speciall kindes of knowledge are comprehended vnder this Science. These foure, Astronomie, Astrologie, Geographie, and Chorographie.[29]

While published after Marlowe's own schooling, Blundeville's text gives an insight into the prevailing trends relating to geographical science in Elizabethan pedagogy. Marlowe was exposed to the subject of 'cosmographie' – and the multitude of endeavours subsumed under its designation – from the beginning. His education, which included a scholarship at King's School in his home town and an intermittent student career at Corpus Christi College, Cambridge, afforded access to geographical texts and also people and communities whose interest in geography was considerable. '[M]aps were part of both formal and informal education in early modern Europe' notes Lesley Cormack; 'From the grammar schools on, both formal and informal educational systems had some interest in the study of the earth and the cosmos'.[30]

As a consequence, it is likely Marlowe would have had contact with both geographical teaching and also substantial textbooks on the subject. A case in point in this regard is John Gresshop, his headmaster at the King's School. Gresshop possessed one of the largest personal libraries in England in the period, encompassing more than 350 volumes and incorporating 'editions of Ovid, Petrarch, Chaucer, and Boccaccio; [...] [and] the comedies of Plautus and the Neoplatonic philosophy of Ficino'[31] as well as numerous geographical texts. The importance of Gresshop's archive and the volumes contained within it for the education of the young Marlowe has been stressed by Vivien Thomas and William Tydeman.[32] Thomas and Tydeman point to the influence of Marlowe's early contact with plays on his own work, in particular the drama he would have read (Terence and Plautus) and observed (religious morality plays such as Everyman, Mankind and The Castle of Perseverance) as a child and adolescent.[33] That many of Marlowe's plays themselves are frequently and consciously historicised and draw on historical figures and myths – the medieval exploits of Tamerlane, the elaboration of the Germanic folk tale of Faustus who sells his soul to the devil, the downfall of Edward II – speaks to the fundamental importance of contemporary source material in understanding Marlowe's work. If, as Thomas and Tydeman claim, Marlowe found his 'knowledge increased' and his 'imagination stimulated' by Gresshop's library, the geographical texts therein can illuminate our understanding of how geography functions within Marlovian dramaturgy.

Perhaps the most notable geographical volume within Gresshop's 'ample' collection, and the first geographically inclined treatise the young Marlowe would likely have encountered, was *Cosmographie* (originally published in 1544), by the influential German cartographer, cosmographer and scholar Sebastian Münster.[34] Münster's role in the development of geography in Renaissance Europe is substantial – according to Benjamin Weiss, Münster 'finally provides a clear link between the study of [Claudius Ptolemy's] *Geography* in an astronomical context and the making of maps'.[35] *Cosmographie*, a multiple-edition work that was constantly revised and augmented throughout the 1500s, was a text bristling with geographical information.[36] If Marlowe was familiar with this storehouse of geographical knowledge from an early age, contemporaneous accounts of its reading suggest *Cosmographie* would have left a deep impression on the imagination of the emerging playwright. The preface to Richard Eden's *A briefe collection and compendious extract of the straunge and memorable things, gathered oute of the cosmographye of Sebastian Munster* (1553), for example, gives an insight into the pleasure induced in the early modern reader by Münster's writings:

> The worke of it selfe is not greate but the examples and varieties are mani so that in a short and smal time, the reader may wander through

out the whole world, and fil his head with many strange and memorable things, he may note the straunge properties of diverse Beastes, Fowles, and Fishes, & the description of far countries, the wonderfull example of sundrye men, and straunge rytes and lawes of far distante nacions.[37]

The *Cosmographie* would have presented to the young Marlowe an admixture of Ptolemaic and other cartographical practices.[38] The promise that the reader 'may wander throughout the whole world' in reading the book resonates with the capability of vicarious travel proffered by early modern cartographers and their maps. Such aspects of sixteenth-century geographical understanding, as we shall see, figure heavily in Marlowe's plays, especially in scenes which explore either overtly or tangentially contemporary geographical science such as the map-reading sequence of the second part of *Tamburlaine*.

Münster's text also incorporated supposedly ethnographic illustrations renderings of fantastical humans, including one-footed giants, double-headed children and lycanthropic wolf-men. Drawn from the fourteenth-century travel narrative *The Travels of Sir John Mandeville*, this extravagant aspect of sixteenth-century geography and its co-existence with more sober mathematical principles would have engendered in Marlowe an awareness of the imaginative possibility of the world describing.[39] The fabulist elements of Münster's presentation of cosmography reveal the possibility made available by placing of imaginative geographies alongside more restrained empirical science. Michael Neill's recognition of Marlowe's 'intoxicated exoticism' is foreshadowed in many regards by the *Cosmographie* of the headmaster's library: just as Münster the cosmographer was renowned for presenting 'the description of far countries, the wonderfull example of sundrye men, and straunge rytes and lawes of far distante nacions' so the Marlovian stage was distinctive for its range of places and diversity of peoples.

If Münster revealed to the school-going Marlowe the imaginative possibilities of cosmography, what can we discern from his experience of geography in his later education? Marlowe attended Corpus Christi College from 1580 to 1587, taking a BA and later an MA, famously breaking his study to engage in 'matters touching benefits of his country', activities that possibly included spying.[40] The Cambridge, the playwright, would have encountered during the 1580s, was an institution abuzz with geographical interest. Geographical textbooks were highly prized by the university libraries. The teaching of geographical subjects was increasingly incorporated into the syllabus at the behest of the scholarly community as a whole. Private map ownership among students themselves increased. The evolving curriculum at Cambridge, the makeup of both the student and teaching faculty, the catalogues of its library, and the private libraries of those whom Marlowe would have encountered during his study there all indicate an environment infused with a flourishing interest in geographical science and its associated subjects such as chorography, astronomy, astrology and cartography.

Cambridge itself was extensively mapped in the latter half of the seventeenth century: included in Christopher Saxton's *Atlas of the Counties of England and Wales* (1579), it was also the subject of several specific surveys, such as the highly detailed town map engraved by Richard Lyne (1574) which incorporates the university.[41] The projection of the county in John Speed's *Theatre of the Empire of Great Britaine* (1611) has been labelled as 'one of [Speed's] finest', attesting to the prominence of the shire in the early modern English cartographic consciousness.[42]

The presence of cartographic activity – surveying, plotting and map-making – would have been reinforced by the everyday exertions of the curriculum. Marlowe's degrees were in the arts, yet both academic qualifications required a level of geographical learning. According to Mark Curtis, the English universities in the second half of the sixteenth century saw a 'broadening and expansion of the arts course'.[43] Such development encompassed geographical science. Tellingly, both teachers and students were the motivators of this disciplinary absorption, underlining the developing popularity of geography or 'cosmography' across the spectrum of the university. '[T]he good will of the tutors and the interest of the scholars', Curtis writes:

> were all that were needed to introduce the study of modern as well as classical history, modern languages as well as Latin and Greek, geography, cosmography, and navigation as well as astronomy, the study of practical politics as well as moral philosophy, and the cultivation of manners, courtesy, and other social graces as well as piety.[44]

As this process indicates, university teachers and scholars began to regard geography and its ancillary disciplines as a central part of educational custom, recognising its inherent benefits to a wide range of professions. 'Geography', asserts Cormack, 'was [...] encouraged and studied by serious students [at Cambridge] following the curriculum, whether they planned a career in the church, in academe, or elsewhere'.[45] Regarding the geographical textbooks Marlowe would have encountered at university, the playwright's MA degree included the study of cosmography and incorporated such influential works as Strabo's seventeen-volume *Geographica*, Ptolemy's *Geographia*, Münster's *Cosmographie* and André de Thevet's *Universal Geography*. Included in this reading list was also William Cuningham's *Cosmographical Glasse*, a text whose characterisation of the delights of 'travelling by map' echoes that of Eden's reception of Münster.[46]

Geographical study at Cambridge

The growing intellectual regard for geography within English university statutes, and specifically at Cambridge, is especially highlighted by the close attention paid to the keeping of the storehouses of such knowledge, geographical books. In 1574, the University's library contained 435 volumes and, according to J.C.T. Oates' retrospective reconstruction from contemporary catalogues, incorporated a separate section or 'stall' designated 'Cosmographia'.[47] Cartographic documents, and instruments, were also highly prized, indicating their value to the scholarly community: in 1582, a document entitled 'Articles for the Office of Keeping the Universitie Librarie' was produced, encompassing an inventory 'conteyning the names of all the bookes and ye number of leaves of all written bookes' in the university library. According to Oates, included in this document is the instruction to the Library-Keeper John Matthew that '"all other bookes of Imagerie with colors, all globes Astrolobes and all other instruments mathematicall, with all other bookes mathematicall or historicall (such as shalbe thought meete by the vicechancellor)" were to be locked up under two keys, of which the Vice-Chancellor was to hold one and the Library-Keeper the other'.[48]

This dictum underscores the care taken by the university libraries of geographic, and in particular cartographic texts and, by implication, their valued status within the library and the university learning space as a whole. '[B]ookes of imagerie with colors' included atlases, bounded collections of maps and texts containing cartographic representations – continental atlases like Peter Apian's *Cosmographicus Liber* (1524) and Abraham Ortelius's *Theatrum Orbis Terrarum* (1570), and also English works such as Saxton's aforementioned atlas of England and Wales (1579) and the antiquarian William Camden's *Britannia*, a predominantly chorographical text but one which also incorporates several maps (1586).[49]

Running parallel to this burgeoning enthusiasm for geography was the increased level of private map ownership among Marlowe's fellow students. In an extensive study of geographical knowledge at English universities in the period, Lesley Cormack writes that 'many Cambridge colleges showed an interest in geography'.[50] Corpus Christi, Marlowe's college at Cambridge, was particularly prominent in this trend. 'The analysis of books owned by students and masters at Oxford and Cambridge, as well as our knowledge of the lives of men interested in geographical topics' notes Cormack, 'shows that some of college and foundations provided special encouragement for the pursuit of geographical studies'.[51] Corpus Christi is listed among the 'best-known loci of geographical interest', emblematic of the wider interest in geography as a subject of study in the universities of late sixteenth-century England.[52] As such, the extent of the private ownership of geographical texts at Marlowe's university during the 1580s 'indicates a genuine and extended interest in the subject'.[53]

A supplementary testament to the vibrancy of geographical learning at Marlowe's college in the second half of the sixteenth century is the geographical literature published by Cambridge graduates and teachers in the period. The Cantabrigian William Cuningham's *Cosmographical Glasse* was a seminal text in the growth of early modern English geographical sciences.[54] His occasional associate John Dee, a former student at St. John's

College, also produced several tracts on astronomy in the 1550s.[55] Figures such as Richard Eden, Christopher Saxton, Thomas Nicholls, and later Thomas Hood exemplify the emerging prominence of geography and its study within the university.[56]

Having mapped out the variety and scope of Marlowe's experience of geographical learning and the vibrancy of 'cosmographical' discourse surrounding his early life and education, I would now like to consider two particular geographical themes described in his drama: the notion of vicarious travel by a map; and the conceptualisation of geographical learning as a hybridised theory/practice intellectual endeavour. Such characteristics reveal Marlowe's deep familiarity with the mapping enthusiasms of the age. Moreover, Marlowe's mapping moments show vividly how such fascinations could be dramatically staged.

'Give me a map, then, let me see'

Marlowe is one of the very few early modern English playwrights to portray the practice of map-reading on stage, despite the popularity of the activity in the period.[57] Nonetheless, he powerfully dramatises the potentialities of cartography as a representative medium. In *Tamburlaine,* in particular, Marlowe's extensive geographical education and knowledge manifest a staging of the map-as-world paradigm that permeated the popular consciousness of contemporary map enthusiasts. The eponymous Scythian shepherd-turned-general summons a map before proceeding to summarise the previous five acts and forecasting the future of his empire:

> Give me a map, then, let me see how much
> Is left for me to conquer all the world,
> That these boys may finish all my wants.
> (5.3.123–125)[58]

The introduction of the cartographic document is mainly motivated by the general's desire to display his military achievements. The audience is taken through Tamburlaine's military campaigns from Persia to Alexandria to Zanzibar to Greece in a striking pre-echo to the colonial practicalities later encountered in the cartographies of Edmund Spenser. It is also driven by a desire to leave a legacy – the general, struck by a mysterious malady, sets out his plans for his successors. Gazing enviously at the regions of the world map as yet unconquered, Tamburlaine concludes by lamenting 'all the golden mines | Inestimable drugs, and precious stones' (151–157) that are out of his reach.

The introduction of the map then serves several functions, including travel narration, grandiose display and military bequest. To serve these functional multiplicities, the play enacts a rhetorical travelling by map – the efficacy of Tamburlaine's oratory relies upon a mutual comprehension of the mapped world he addresses. The enumeration of place names, the frequent geographical positioning and the indication of distances amalgamates the map-reading with a travel narrative, recreating a sense of travelling through the world. In doing so, Marlowe, the playwright, evinces an awareness of the intellectual belief in the map as a medium for vicarious travel, as celebrated in the works of Elyot, Cuningham, Norden, Burton, Donne, Middleton and Herrick.

The heightened drama of the map moment corresponds with Tamburlaine's heightened dramatic sensibilities. Provoking a cessation of action in a scene that has witnessed the protagonist order his troops to such heights of derring-do as 'march against the powers of heaven | And set black streamers against the firmament | To signify the slaughter of the gods' (5.3.47–50) and 'Strike the drums, and, in revenge of this,| Come, let us charge our spears and pierce his breast | Whose shoulders bear the axis of the world' (5.3.57–59), Tamburlaine's command, 'Give me a map', has an arresting effect within the dynamic of the play, drawing all eyes to the general and his actions. The protagonist, as readers have noted, revels in the theatricality of his actions and their symbolic suggestions, for example, the coercive yoking of kings to his chariot (5.1.147) or the atheistic burning of the Koran (5.1.176–200).[59] That an instance of Tamburlaine's enthusiasm for ostentatious demonstration involves cartography highlights Marlowe's own cartographic enthusiasm, elevating the map as a prop to an object pregnant with symbolic meaning.

Thus, in utilising the map as a theatrical appurtenance and giving it such a prominent role in the conclusion of Tamburlaine's chronicle, Marlowe demonstrates an attraction to the cartographic document familiar among his contemporaries. As a physical artefact, the map proved a persistently fascinating object for early modern intellectuals: 'What greater pleasure can there now be', asks Marlowe's near-contemporary Robert Burton, 'than to view those elaborate maps of Ortelius, Mercator, Hondius, etc?'.[60] Ethel Seaton has argued that the scene indicates the playwright's alignment with the wider enthusiasm among thinkers in the Renaissance for cartography. Seaton highlights, in particular, the conspicuous parallels between Marlowe and the cartographic enthusiast Burton. Citing the extract from Burton's *Anatomy of Melancholy* quoted above, her main intention is to identify the similarities between Tamburlaine's extensive geography – in scale, dramatic progression, description and locational nomenclature – and Abraham Ortelius's *Theatrum Orbis Terrarum*.[61]

However, Seaton's comparison between Tamburlaine's cartographic action and Burton's map-reading activities also highlights the similarities between Marlowe's understanding of the map and Burton's own conceptualisation of cartographic study. As noted, Burton's encomium reiterates established notions surrounding cartography and its study. In staging a map-reading, Marlowe instantiates the 'extraordinary delight' of map study, presenting dramatically a link between earlier thinkers such as Elyot and Cuningham and their seventeenth-century counterparts like Burton, who espoused faith in the cartography's unique representational mode.

The visual power of cartography

This unique representational mode stems from the map's perceived aptitude to go beyond the immediately visible and project a multi-dimensional space within the bounds of the stage, as it does in the study. The dramaturgical effectiveness of Tamburlaine's rhetoric depends upon the ability of the addressed lieutenants and the audience to view on the map a three-dimensional space teeming with variety. In Tamburlaine's directive 'let me see how much | Is left for me to conquer all the world', the emphasis on the visual capacity of the map to facilitate an ocular comprehension of 'all the world' reflects the map enthusiast's passion for cartography's variety. In Tamburlaine's order and its presumption of the map's ability to portray the world in its entirety, we can perceive an echo of William Cuningham's instruction to his reader to look on the 'cosmographical glass' and 'beholde the diuersitie of countries: natures of people, & innumerable formes of Beastes, Foules, Fishes, Trees, Frutes, Stremes, & Meatalles' in the map'.[62] Both texts inscribe a language of totality where the microcosmic simulacrum is transformed into the macrocosmic entirety.

The rhetorical function of Tamburlaine's language is further indicated by its diversity. The distinguishing facet of the journey by the map is a stress on cartography's condensation of huge expanses into singularly perceivable vistas. While emphasising multiple locations, vast distances and topographical diversity, Marlowe also condenses the vastness of Tamburlaine's journeying into a single speech. In barely 34 lines of iambic pentameter, Tamburlaine cites 21 separate toponyms.[63] These vary dramatically in topographical identity, encompassing lands, seas and lakes, recalling Cuningham's 'diuerse' taxonomy – for example, in the first few lines of his speech, the general begins by naming Persia (126), then the Caspian Sea (127) and afterward Egypt, Arabia, Alexandria (130–131) before the conflux of the Mediterranean ('the Terrene') and the Red Sea (133). The assortment of locations and geographical features is notably interspersed with Tamburlaine's narrative, articulated through a description of his own action. Both journey and empire are contracted and condensed into a single speech, much like the mapmaker condenses the world in a single sheet. In the first 14 lines of his speech, Tamburlaine announces:

> Here I began to march towards Persia,
> Along Armenia and the Caspian Sea,
> And thence unto Bithynia, where I took
> The Turk and his great empress prisoners.
> Then march'd I into Egypt and Arabia;
> And here, not far from Alexandria,

Whereas the Terrene and the Red Sea meet,
Being distant less than full a hundred leagues,
I meant to cut a channel to them both,
That men might quickly sail to India.
From thence to Nubia near Borno-lake,
And so along the Aethiopian sea,
Cutting the tropic line of Capricorn,
I conquer'd all as far as Zanzibar.
(5.3.126–139)

Throughout his reading of the map, Tamburlaine is both stationary and on the move. He is characterised as a travel guide, with the general taking his lieutenants and the audience on tour through the expanses of his empire via the medium of the cartographic document. It is the maintenance of this apparent paradox – of travelling and remaining stationary – that indicates the protagonist's belief in the map's ability to surpass the restrictions of his own physical horizons. Parallels with the contradictory nature of Cuningham are discernible. The reader of *The Cosmographical Glasse*, Cuningham insists, is able to 'beholde the diuersitie of countries' and their accompanying topographies, eco-systems and ethno-geographical attributes while remaining within the confines of the study.[64] Similarly, Tamburlaine will survey his empire through the act of map study and through the medium of the map. For Renaissance map enthusiasts, map-reading was understood to provide a unique substitute for the sensations of actual travel. Tamburlaine enacts a vicarious journey through the map, again anticipating other Renaissance intellectuals who travelled by map, including Burton: 'I neuer trauelled but in a Map or Card'.[65]

  Further underling, these elements – the proliferation of place names, the range of geographical locations, Tamburlaine's verbal movement through the cartographic document – and buttressing their evocation of the world of the map is the constant reference to distance. This functions initially to emphasise the physical vastness of Tamburlaine's conquests:

Then by the northern part of Africa,
I came at last to Greacia, and from thence
To Asia, where I stay against my will,
Which is from Scythia, where I first began,
Backward and forwards, near five thousand leagues.
(140–144)

Tamburlaine foregrounds the spatial puzzle of maps as 'worlds on worlds', where the single two-dimensional vista can contain a multi-dimensional 'world' with panoramic views. Alongside a recurrence of the traversing travel narrative and its adverbial phrasing ('backward and forwards'), the contradiction of perceiving vast spaces, variegated geographies and multiple vistas in a small pictorial representation and the single view is over-come by Marlowe's protagonist and his belief in the map's ability to exceed the boundaries of the real.

The map as will

Tamburlaine's belief in what Donne describes as the 'world on world' is developed as the scene progresses. Lisa Hopkins astutely points out that Marlowe's play is clearly intensely interested in the visual both in terms of display and of emblematic force and that 'Tamburlaine is never happier than when he is orchestrating a dramatic self-staging':[66]

> For him, theatre is a fundamental tool of power, and one of the reasons why he was both so revelatory as a hero in his own day and so memo-rable as one in ours is undoubtedly the fact that he is conceived of as so essentially theatrical a character.[67]

This sense of 'dramatic self-staging', of theatre as a 'fundamental tool of power', and the heightened theatricality of Tamburlaine as a character is exemplified in the appearance of the map and the subsequent speech. Initially, the map is summoned for the general's own use, that he might survey the bounds both of his empire and also that which has remained unconquered. Yet, this statement of personal utility is soon overwhelmed by the protagonist's awareness of his imminent demise: sensing that his own 'martial strength is spent', the general reconfigures the purpose of the map from the personal to the public. The line immediately preceding the stage direction that indicates the appearance of the map – 'That these my boys may finish all my wants' – marks a subtle shift in the function of the cartographic document as a theatrical prop. Changed from an object of personal contemplation and reflection to a medium of performance and bequest, the map is forcefully subsumed into Tamburlaine's attempt to bequeath a military legacy.

However, despite this shift of purpose, the conceptual belief in the car-tographic world pervades Tamburlaine's oratory. Such is the permanence of the idea of the map as the world in the protagonist's mind, he assumes a similar belief in not only the audience but also the rest of the characters onstage. Fundamental to the efficacy of his self-aggrandisement, egotistic celebration of his soldierly brilliance and forceful expression of his military will – all contained within the speech immediately following the appearance of the map (5.3.126–160) – is the conveyance to his lieutenants of the scope of his empire. Such a conveyance draws much of its force from the presence of the cartographic document. The rhetorical efficacy of the speech is pred-icated largely upon the notion that, like their general, Tamburlaine's 'boys' can perceive in the map not simply a two-dimensional document. They can also go beyond the constraints of 'real' physical perception to survey the world itself.

The map-reading sequence in Tamburlaine exemplifies Marlowe's awareness of contemporary atlases such as the Mercator. It also gestures towards the prevailing enthusiasm for cartographic study. Furthermore, through the rhetoric of the protagonist, it reveals the playwright's sensitivity to the concept of the world of the map. Tamburlaine's speech articulates the Renaissance faith in the paradigm of the 'world on world' and its many facets – condensation of space, affording of panoramic vistas, the opportunity for safe travel, atemporal traversing of vast landscapes. Tamburlaine's command also draws the eyes of all in the play, and in the audience, towards the cartographic document. In this sense, the scene stages a major moment in the history of early modern cartography – the monosyllabic order 'Give me a map' commences a study of cartography in the early modern English theatre.[68] Consequently, in bringing the map onstage and into direct contact with the space of the theatre and its audience's mass gaze, Marlowe stages the full dramatic power inherent in contemporary understandings of the map's representational potential.[69]

Proving cosmography

If *Tamburlaine* stages the world-as-map concept to lend a dramatic force to the conclusion of his play, *Doctor Faustus* sees Marlowe turning his sensitivity to prevailing geographical concepts onto the cosmographer. In particular, Marlowe explores the tension between bookish learning and practical experience within the discipline. The rhetorical actions of the titular protagonist in his attempts to 'prove cosmography' are subtly undercut and his efforts to attain divine knowledge through the discipline – a crucial aspect of cosmography – remain unfulfilled. Where *Tamburlaine* demonstrates Marlowe's consciousness of cartography as a means of imaginative travel, *Faustus* signals the playwright's receptiveness to the theoretical understanding of geographical science as a discipline encompassing both theory and practice. Where Tamburlaine travels by map, it is through real and not vicarious travel that Faustus ultimately achieves his own comprehensive geographical and cosmographical erudition.

For early modern playgoers familiar with the Marlovian partiality for travelling, Faustus would have been initially startling in his inertia. At the play's opening, the protagonist is seen in his study at the University of Wittenberg spending his time among his books (1.1.13–15). His initial situation diverges from that of the continent traversing of Tamburlaine. Faustus is, tellingly, characterised by inaction: at first glance, Faustus appears strikingly contained. The introduction of the main character in *Tamburlaine* figures the general as a perennial agent of action through the use of present tense:

> [...] the Scythian Tamburlaine
> Threat'ning the world with high astounding terms
> And scourging kingdoms with his conquering sword.
>
> (1. Prologue. 4–6)

The appurtenances of Tamburlaine – in this instance, his 'conquering' sword – are imbued with the immediacy of action characteristic of its owner. However, in *Faustus* there is in the opening scenes a conspicuous contrast to the active *in medias res* beginning of *Tamburlaine*. The chorus to Faustus introduces the 'tragical history' and the 'life and death' of its protagonist with a concerted contraction of space, environment and action:

> Not marching now in fields of Trasimene,
> Where Mars did mate the Carthaginians;
> Nor sporting in the dalliance of love,
> In courts of kings where state is overturn'd;
> Nor in the pomp of proud audacious deeds,
> Intends our Muse to vaunt her heavenly verse:
> [...] Now is [Faustus] born, his parents base of stock,
> In Germany, within a town call'd Rhodes:
> Of riper years, to Wertenberg he went,
> Whereas his kinsmen chiefly brought him up.
> [...] And this the man that in his study sits.
>   (1. Prologue. 1–6, 11–14, 28)

Verbalised in negatives (not, nor, now), Faustus appears withdrawn, hermitic and stationary. The contrast with the martial vitality of Tamburlaine is unambiguously outlined by the first two lines, which evoke the Roman God of War, military conflict and movement as a contrast to the opening situation of the play – 'Not marching now in fields of Trasimene | Where Mars did mate the Carthaginians'. Finally, the conclusion underlines Faustus's fixedness: 'And this the man that in his study sits'.

Nevertheless, notions of travel and, by implication, an interest in geographical variety are prominent from the outset of the play. While Faustus may be cloistered, the sense of travel is never far from centre stage. Marlowe describes his protagonist in the following fashion: 'So soon he profits in divinity | The fruitful plot of scholarism graced' (1.1.16). Ostensibly narrating Faustus's elevation to erudition, these lines also amalgamate real and mental space, striking a plangent note with contemporary geographical discourse. The 'plot of scholarism graced' by Faustus may be glossed in two distinct yet connected ways – firstly, the physical surrounds of the university and also the cerebral mindscape of learning. Even in the restrained area of academic study, Marlowe's inclination towards movement is apparent.

Additionally, the characterisation of Faustus suggestively invokes classical myth, pushing to the centre of the stage the theme of travel:

> [...] shortly he was graced with doctor's name,
> Excelling all, and sweetly can dispute
> In th'heavenly matters of theology;
> Till, swoll'n with cunning of a self-conceit,

His waxen wings did mount above his reach,
And, melting, heavens conspired his overthrow.
For, falling to a devilish exercise,
And glutted now with learning's golden gifts,
He surfeits upon cursèd necromancy.

(1. Prologue. 17–25)

The comparison of Faustus to Icarus instantly evokes a tale which contains at its very heart an (ill-fated) trip. Such recourse to classical mythology and its foregrounding of travel is given especial significance in the context of Marlowe's will to movement conspicuous elsewhere in the playwright's works. In *Edward II*, the drama opens with geographical movement, or rather the appeal to geographical movement. In a letter to his 'favourite' Gaveston, the newly crowned monarch Edward pleads '["]come [...] | And share the kingdom with thy dearest friend"'(1.1–2). Edward's request is infused with implorations to travel:

'My father is deceas'd! Come, Gaveston,
And share the kingdom with thy dearest friend,'
Ah! words that make me surfeit with delight!
What greater bliss can hap to Gaveston
Than live and be the favourite of a king!
Sweet prince, I come; these, these thy amorous lines
Might have enforc'd me to have swum from France,
And, like Leander, gasp'd upon the sand,
So thou would'st smile, and take me in thine arms.
The sight of London to my exil'd eyes
Is as Elysium to a new-come soul;
Not that I love the city, or the men,
But that it harbours him I hold so dear—
The king, upon whose bosom let me die,
And with the world be still at enmity.

(1.1. 1–15)

Recalling the Icarus metaphor for Faustus, Gaveston imagines himself as 'like Leander', evoking the latter's nightly swim across the Dardanelles. Such characterisations exemplify a persistent theme of Marlowe's writing, namely the adoption and adaptation of Greco-Roman mythological traditions in his own work. As Sara Munson Deats notes, 'Among the poetic manipulators of the classical kaleidoscope, few handled these many-faceted images with more skill, assurance, and complexity than Christopher Marlowe, translator of Ovid and Lucan, adapter of Ovid and Virgil'.[70] Consequently, the figuring of Faustus as a traveller is a carefully chosen and meaningful move, foregrounding travel from the outset of the play.

Faustus the cosmographer

While the introduction of Faustus foreshadows a travelling protagonist and a travelling drama through its focus on itinerancy, Marlowe's work engages with geography in a number of other expressive ways. A case in point is the geographically inflected concerns running through his contract with Mephistopheles, where he exchanges his soul for preternatural knowledge and abilities. In many ways, these demonstrate the central character's striving to assume the role of cartographer of his own self-serving geography. Revealingly the first of Faustus's demands in exchange for his soul is ease of journeying: he desires the ability to transcend the bounds of corporeality and travel freely and without duress.

> Then there's enough for a thousand souls.
> Here, Mephistopheles, receive this scroll,
> A deed of gift of body and of soul -
> But yet conditionally that thou perform
> All articles prescrib'd between us both.
> Mephistopheles: Faustus, I swear by hell and Lucifer
> To effect all promises between us made!
> Faustus: Then hear me read them.
> 'On these conditions following:
> First, that Faustus may be a spirit in form and substance'.
> (1.1. 89–97)

Faustus's contractual condition of 'being a spirit in form and substance' and assuming a form not bound by physical reality is anticipated earlier in the play when, ruminating on the possibilities afforded by a potential pact with the devil, he announces:

> I'll be a great emperor of the world
> And make a bridge through the moving air,
> To pass the ocean with a band of men;
> I'll join the hills that bind the Afric shore,
> And make that country continent with Spain,
> And both contributory to my crown.
> (1.3. 105–110)

The construction of a 'bridge through the moving air, | To pass an ocean with a band of men' reveals the motivation behind the desire to 'be a spirit in form and substance'. Faustus's intention is to circumvent the practicalities and hardships of travelling which 'real' geography necessitates. Such an objective bears a striking similarity to those map-readers in the period like Burton, who enjoyed travel by card, chart or globe. In addition, the ability to travel is supplemented by a more fundamental aptitude – the capability to alter the very topography of the land to facilitate his desires. The 'bridge through the moving air' in particular resolves the difficulties of the military expedition by sea, enabling Faustus to potentially adopt the conqueror-like status of his Marlovian predecessor.

In articulating such a desire, Faustus also exhibits his impulse to fashion in a similar way to cartography, his own schema of the existing landscape. Such actions, driven by an egotistic attempt to increase his power, echoes Tamburlaine's bombastic ambition to rewrite the map of the world and confound prevailing geographical understanding.[71] John Gillies has perceptively observed that Tamburlaine's remark interrogates the very notion of 'geography' as a single entity. Tamburlaine, Gillies argues, 'proposes a distinction between "geo-graphy" as a world creating act of graphic depiction, and "geography" as a depicted world-object'.[72] Gillies's description of Tamburlaine's proposition chimes with Faustus's language. His request to Mephistopheles is marked by his arrogation of geography in the expression of power. The purpose of Faustus's attempt to manipulate the continent of Europe conveys the central character's cognizance of the power of geographical control. The manipulation of the topography of Europe will not only permit personal ease of travel for him; but it will also facilitate a movement of a more militaristic bent. He will travel accompanied by a 'band of men' simultaneously expanding and occupying those areas 'contributory to my crown'. Marlowe fashions Faustus as a cartographer appropriating an assumed ability to re-inscribe the world's topography for personal ends. The closing up of the Mediterranean, an act of geographical manipulation, is motivated by the desire to integrate both Africa and Europe into one entity and make 'both contributory to my crown'.

Exploring and inscribing the earth

If the geographically minded demands of Faustus's deal with the devil imply a geographical subtext, later developments in the play show Marlowe staging one of the primary tenets of early modern geographical thought: the necessity of achieving a synergy between theory and practice. While he is initially characterised as a traveller whose 'waxen wings did mount above his reach' (1. Prologue. 21), Faustus's cosmographical endeavours, in particular, underline Marlowe's responsiveness to the importance of the practicalities of travel in geographical learning. In doing so, it displays an intellectual kinship with geographers such as Walter Raleigh, who sought to demonstrate the propinquity and even coincidence of world-describing and exploration.

Emblematic of this trend is the sub-discipline of 'mathematical geography'. One of the most substantial topics covered in the study of geography in the early modern period, mathematical geography, was, according to Lesley Cormack, also 'inexorably linked with practical application, even in its most arcane manifestations'.[73] Within cosmography, theoretical knowledge sought practical substantiation – Cormack cites the importance of mapping the globe for the burgeoning British imperial enterprise as illustrative of how 'mathematical geographers helped to develop a new methodology that combined mathematics, inductive experimentation and practical application'.[74] But other efforts such as the exploration of the eastern sea-board of America and attempts to discover the Northwest Passage, also foregrounded a symbiotic relationship between knowledge and application.

While certain explorers and privateers were renowned as men of action and ignorant of bookish learning, others were deeply immersed in the intricacies of cosmographical learning.[75] In his *History of the World* (first published in 1614), Walter Raleigh, who alongside Frobisher was one of the most prominent English navigators of the period in which Marlowe flourished as a playwright, displays a learned awareness of cosmography and its associated disciplines of cartography, astronomy, hydrography, astronomy and mathematical navigation. Maps are included in the voluminous text, alongside an acknowledgement of the conventions of mapping. Introducing his work in the preface, Raleigh references cartographical representation of topographical features: 'And herein I haue followed the best Geographers: who seldome giue names to those small brookes, whereof many, ioyned together, make great Riuers; till such time as they become vnited, and runne in a maine streame to the Ocean Sea'.[76] Likewise, Raleigh's belief in the importance of participating in the development of geographical knowledge through his own surveying is apparent:

> but because the Land of Canaan, and the borders thereof, were the Stages and Theaters, whereon the greatest part of the Storie past, with that which followeth, hath beene acted, I thinke it very pertinent (for the better vnderstanding of both) to make a Geographicall description of those Regions: that all things therein performed by the places knowne, may the better bee vnderstood, and conceiued.[77]

Raleigh figures the importance of 'geographicall description' as a means of enhancing the understanding and conceptualising biblical history and events recorded in scripture. Geographical representation or 'description' is figured as an exegetical tool to construct 'the Stages and Theaters, whereon the greatest part of the Storie past, with that which followeth, hath beene acted'. Raleigh's tract is striking because it exemplifies the overlapping of the theatre ('Stages and Theaters') and the geographical space; and the articulation of the close symbiotic relationship between exploration and surveying in the late-sixteenth and early seventeenth century. The success of cosmographical understanding is dependent on a collaborative effort between both the 'Storie past' and 'that which followeth'.

Such an idea, with its emphasis on both theoretical and experiential learning in geographical science, features heavily in Marlowe's play. David McInnis, noting Peter Holland's description of Marlowe's drama as 'the greatest of all Elizabethan plays of travel',[78] has argued that its cosmographical development 'anticipates [Francis] Bacon's emphasis on empiric validation'.[79] Accordingly:

> When Faustus [...] exhausts the possibilities for mentally testing the knowledge recorded in his books, he is inspired [...] to see the things he reads about, to verify their wonder first-hand where abstract rules cannot yield the verification he craves. He becomes a learned traveller [...], recognising and appreciating the sites he sees, armed with fore-knowledge provided by his books, and 'rauished with delight' at the spectacles.[80]

McInnis foregrounds the interchange between theory and practice in early modern cosmography, characterising Faustus as an empiricist whose geographic efforts are enhanced by exhaustive bookish learning and subsequent travel. In his play, Marlowe expertly explores this symbiosis, ironically presenting a failed cosmography-by-book and heroically depicting a successful cosmography-by-experience. He is a proto-Baconian, a forerunner of the Royal Society whose symbolic success emerges in the pursuit of geographical knowledge.

To do this, Marlowe first establishes the inadequacy of bookish learning. Faustus's initial exploration of cosmography begins with his interactions with Mephistopheles. Importantly for evaluations of Faustus's depiction of cosmography as an intellectual discipline, the emergence of the 'studious artisan' (1.1.57) corresponds with an attempt to establish a comprehension of theological geography.[81] However, the interaction between Faustus and Mephistopheles portrays the insufficiency of rhetorical investigation alone in the pursuit of cosmographical inquiry. Faustus attempts to discover 'the structure and motions of the heavens', asking, 'Come, Mephistopheles, let us dispute again | And argue of divine astrology' (2.3.33–34). The exhortation by Faustus echoes the academic experience of Cambridge during Marlowe's time.[82] 'To many', writes Jerry Brotton, 'the cosmographer seemed to adopt a divine perspective from which to gaze upon the earth, while also looking up and speculating on the structure and origins of the universe'.[83] In commencing his astrological discussion, Faustus repeats the popularly understood concern of the mapper or 'cosmographer': discernment of a godly agency.

This persistent striving for an understanding of God through mathematical calibrations of the earth and the cosmos is replicated by Marlowe and is evident in Faustus's first forays into the discipline and its methods:

> Tell me, are there many heavens above the moon
> Are all celestial bodies but one globe,
> As is the substance of this centric earth? [...]
> But, tell me, have they all one motion, both situ et tempore? [...]

> But, tell me, hath every sphere a dominion or intelligentia? [...]
> How many heavens or spheres are there? [...]
> Well, resolve me in this question; why have we not conjunctions, oppositions, aspects, eclipses, all at one time, but in some years we have more, in some less?
>
> (2.3.35–38, 44, 54–56, 58, 61–63)

Faustus's intensive interrogation of his demonic servant parallels Blundeville's description of the pursuit of astrology as part of cosmographical inquiry. Answering the question 'What is Astrologie?' Blundeville writes: 'It is a Science which by considering the motions, aspects, and influences of the starres, doth foresee and prognosticate things to come'.[84] Faustus's questions on the motion, substance and number of celestial bodies parallel Blundeville's considerations regarding the structure of the universe. The discussion with Mephistopheles may be characterised as, in part, a cosmographical discourse, with Faustus, the budding cosmographer attempting to discover 'motions, aspects, and influences of the starres' and 'prognosticate things to come'.

However, in an intriguing twist, Marlowe destabilises any attempt to push cosmographical enquiry to its ultimate goal, namely the detection of the divine. The linguistic intertwining of astrology, astronomy and theological divination by Faustus in his request to 'dispute again | And argue of divine astrology' is not an exclusively Marlovian construct. The discipline was imbued with a religious purpose, serving as a means to discern the 'structure and origins' of the cosmos.

It is precisely this attempt to discern origin that evades Faustus at the last moment. Contemporaneous accounts of cosmography and its pursuit validate Brotton's description of the discipline's impulse towards theological understanding. In his preface to the rhetorician and logician Petrus Ramus's *The Way of Geometry* (1636) – a work which discusses cosmography –the English mathematician William Bedwell draws on the authority of Plato to refute the claim that 'Geometry may no way further Divinity, and therefore is no fit study for a Divine':

> Plato saith, That God doth alwayes worke by Geometry, that is, as the wiseman doth interprete it[.] [...] Dispose all things by measure, and number, and weight: Or, as the learned Plutarch speaketh, He adorneth and layeth out all the parts of the world according to rate, proportion, and similitude. Now who, I pray you, understandeth what these termes meane, but he which hath some meane skill in Geometry.[85]

Bedwell's focus on the use of geometry as a means of discerning divine truths is foreshadowed in John Dee's preface to a translation of Euclid (1570). Again, the emphasis is on facts and figures discernible by geometric calculations, which will eventually lead to a greater understanding of God's design. Quoting the Roman philosopher Boethius, Dee writes, 'All thinges (which from the very first originall being of thinges, have bene framed and made) do appeare to be Formed by the reason of Numbers. For this was the principall example or patterne in the minde of the Creator'.[86] That the 'mind of the creator' could be distinguished through an investigation into cosmographical structure depended upon the notion of God as an ordered and mathematically sound creator, a notion which drew its impetus from the humanist effort to assimilate Platonist ideas into the Christian world-view. As Paul H. Kocher writes: 'All that was needed was to make the Christian God the architect of this mathematical universe. The result was the maxim, God geometrizes'. As a consequence, '[f]ew Elizabethan writers on mathematics failed to bring in somewhere [the] idea of God the geometer'.[87] The processes of geometry and mathematical geography are more generally aggregated to produce an image of the cosmos and, as a consequence, God through his creation.

It is within this tradition of theologically modulated mathematical analysis that Faustus's cosmographical investigations operate. While Faustus may be viewed in parts of the play as a paradigm of sensual indulgence, it is clear that in his discussion with Mephistopheles, the protagonist's ultimate aim is theological.[88] The astrological inquiry into the divination of 'things to come' decisively climaxes with Faustus's query as to the originator of the world (2.3.65). Each of the questions regarding substance and temporal oscillations of the earth and stars is answered by his servant in an almost perfunctory way – the iambic pentameter form that persists throughout the play is broken by the unmetrical exchange of technical data:

Faustus: But tell me, have they all one motion, both situ et tempore?
Mephistopheles: All jointly move from east to west in twenty-four hours upon the poles of the world; but differ in their motion upon the poles of the zodiac.

(2.3.44–47)

and:

Faustus: Tush!
These slender trifles Wagner can decide;
Hath Mephistopheles no greater skill?
Who knows not the double motion of the planets?
The first is finish'd in a natural day;
The second thus: as Saturn in thirty years; Jupiter in twelve; Mars in four; the Sun, Venus, and Mercury in a year; the moon in twenty-eight days. Tush, these are freshmen's suppositions. But tell me, hath every sphere a dominion or intelligentia?
Mephistopheles: Ay.

(2.3.48–57)

Yet when Faustus seeks the fundamental question of his cosmographical enquiry – 'Well, I am answered. Tell me who made the world' – the response startles the protagonist into indignation:

> Mephistopheles: I will not.
> Faustus: Sweet Mephistopheles, tell me.
> Mephistopheles: Move me not, for I will not tell thee.
> Faustus: Villain, have I not bound thee to tell me anything?
> Mephistopheles: Ay, that is not against our kingdom; but this is.
> Think thou on hell, Faustus, for thou art damn'd.
> Faustus: Think, Faustus, upon God that made the world.
>
> (2.3.66–72)

At the conclusion of this first attempt at the cosmographical inquiry, Faustus's quest for the fundamental knowledge of the discipline – the origin and originator of the world – is left unanswered. In the interrogation of Mephistopheles, we can discern what Stephen Orgel calls Faustus's 'drama [...] of overreaching ambition combined with relentless failure'[89] refracted through the lens of cosmography. Marlowe incrementally develops the line of enquiry with increasing revelations about 'divine astrology' before finally leaving the main question unanswered. As if to aggregate the ironic effect of Faustus's failed cosmographical investigations, his lack of certainty regarding the ultimate truth he seeks is compounded by the reiteration by Mephistopheles of his damnation: 'Think thou on hell, Faustus, for thou art damn'd'. The unanswered question prompts the protagonist to return and seek solace in the truths of his previous, pre-damned self: 'Think, Faustus, upon God that made the world'.

Practising cosmography: Travelling

If the astrological discussion, with its climactic conclusion, sees Marlowe theatricalise the limits of discursive, theoretical cosmography in its search for knowledge of the divine, the later events of the play turn to foreground practical cosmography. Faustus, having failed to determine divine origins through discourse, achieves his goal through action. In doing so, Marlowe's play reiterates contemporaneous ideas about the discipline: cosmographical investigation in the early modern period was characterised by an investigation of grander questions about the construction, nature and origin of the universe. According to William Cuningham's *Cosmographical Glasse* (1559), the subject was pre-eminent among scientific disciplines:

> If ever ther wer Art for all mens vse inuented, Science set forth wherein consisteth Sapience, or Treasure worthy to be had in estimation: no doughte (louynge Reader) either Cosmographie is the same, or els it

is not to be founde vppon th'Earth. [...] Moreouer mannes helth (without whiche Honour, Fame, Richesse, Frendes, and Life it selfe, semeth bothe troublous, and noysome) can not be conserued in perfite estate, or once lost be recouered and restored without Cosmographie. For howe greatlye herein it profiteth, to consider the temperature of Regions, Cities, and Townes, in what Zone, & vnder what Clymate and Parallele they are situated.⁹⁰

This effusive acclamation of cosmography, and by implication, its practitioners, propagates the pre-eminence of the subject within the wider schema of 'Art [...] invented' and 'Science set forth'. Cuningham also stresses cosmography's 'ample vse' and 'manifold benefites', maintaining the crucial importance to the basic function of human society:

And if I shall begin with the defence of our Country, which ought to be more praecious, the Parentes wife Children or Consanguinitie, Cosmographie herein do so much profite, that without it both valeaunt Corage, Policy and Puisaunce oftentimes can take no place. For by her we are taught whiche way to conduct most safely [...], where to pitch oure tentes, where to winter: yea, and where most aptlye to encounter with them in the fielde.⁹¹

Against this backdrop of an emphasis on the practical benefits of the discipline of cosmography, Faustus attempts to assume mastery. However, as scholars have noted, the drama retains a recurrent emphasis on the superficiality of Faustus's achievements.⁹² The protagonist's ultimate damnation throws into stark relief the asininity of what Andrew Sofer calls Faustus's 'jejune parlour tricks'⁹³ and David Bevington and Eric Rasmussen describe as 'scenes of buffoonish revelling in sin',⁹⁴ as exemplified by the mischief at the court of the pope (3.1.1–96), cuckoldry jokes played on the knight of the Holy Roman Emperor (4.1.65–92), and fornication with a devil masquerading as Helen of Troy (5.1.88–109).

Faustus's attention to geography in his negotiations with Mephistopheles prefigures his ensuing practical cosmographic exertions, especially in scale. As the stationary character of Faustus is displaced in the third act with the tour of Europe, the characteristic trait of expansiveness in Marlowe's plotting comes to the fore with an enactment of experiential geographical science. Where before Faustus's failed exploration of cosmography takes the form of a rhetorical interaction, as the play develops the practical pursuit of 'divine astrology' in the second half of the play proves more prosperous than the academic discussion with Mephistopheles. As the introductory chorus tells us in act three, quitting his study at Wittenburg and venturing south towards Italy, the eponymous scholar embarks on a journey that ranges from 'Olympus top' through the starry firmament before descending to the papal court at Rome:

Learned Faustus
To know the secrets of Astronomy
Graven in the book of Jove's high firmament,
Did mount himself to scale Olympus' top.
Being seated in a chariot burning bright,
Drawn by the strength of yoky dragons' necks,
He now is gone to prove cosmography,
And as I guess will first arrive at Rome,
To see the Pope and manner of his court,
And take some part of holy Peter's feast,
That to this day is highly solemnized.
                              (Chorus. 1–11)

Wagner announces that his master 'is now gone to prove cosmography' (3. Chorus. 7). At once splitting in two knowledge of cosmography, this remark presents to the audience the popular understanding of cosmography as one discrete scholarly subject that simultaneously comprises a binary of bookish learning and inductive experience. The expanded B-text of the play emphasises the full range of Faustus's cosmographical journey:

He views the clouds, the planets, and the stars,
The tropic zones, and quarters of the sky,
From the bright circle of the horned moon
Even to the height of Primum Mobile;
And, whirling round with this circumference,
Within the concave compass of the pole,
From east to west his dragons swiftly glide.
                              (Chorus. 7–13).[95]

The travel guide tone, with its paratactic formulations and progression from Germany to France to Germany and then onto Italy and Rome to the court of the pope, serves to stress the travel narrative of the protagonist. The use of present tense and enumeration buttress the feverish variety of Faustus's actions.

More decisively in the context of the play's delineation of cosmography, the journey establishes a new turn in Faustus's cosmographical inquiry. Faustus aims 'to prove' the 'cosmography' demonstrates Marlowe's awareness of the understood sixteenth-century characterisation of the geographer as both study-bound scholar and fearless explorer. Faustus is initially described as 'grac[ing] the fruitful plot of scholarism', circumscribed by books and academia. Complaining of the limitations of his university learning, he remarks, 'Is to dispute well logic's chiefest end? | Affords this art no greater miracle?' (1.1.8–9). In seeking the 'greater miracle' Faustus is provoked firstly to his interrogation of Mephistopheles. Having proved fruitless in his attempts at discerning the divine, Faustus's characterisation as the ideal cosmographer foreshadows his involvement in practical action. The reference at the opening of the speech to 'the book of Jove's high firmament' is, in this context, telling. Figuring the heavens as a written text wherein is 'graven' the 'secrets of astronomy', Marlowe's lexical choice presents a link back to the bookish learning and disputation that characterised the earlier scenes but also subtly emphasises the importance of the 'book' of the cosmos, read through practical exploration.

Theoretical failure and practical success

Crucially, this attempt to prove cosmography proves immeasurably more successful than any rhetorical discussion with Mephistopheles, imparting a profound understanding of the cosmos and its movements. In the introduction to the fourth act, the chorus announces that the cloistered scholar's cosmographical ambitions have motivated travelling, surveying, and 'prov[ing]' the precepts learned in his books. As a consequence, his knowledge of cosmography has increased exponentially. The chorus describes Faustus's return to Wittenberg and discussions with his 'friends and nearest companions':

> When Faustus had with pleasure ta'en the view
> Of rarest things, and royal courts of kings,
> He stay'd his course, and so returned home;
> Where such as bear his absence but with grief,
> I mean his friends, and near'st companions,
> Did gratulate his safety with kind words,
> And in their conference of what befell,
> Touching his journey through the world and air,
> They put forth questions of Astrology,
> Which Faustus answer'd with such learned skill,
> As they admir'd and wond'red at his wit.
>   (4. Chorus. 1–11)

Francis Dolores Covella has shown how 'the action of the play is framed by the presence of respectable scholars who represent the norm of scholarly behaviour and values'.[96] The knowledge garnered from his travels or 'journey through the world and air' has rendered Faustus pre-eminent among his colleagues to the extent that he is treated as an authority on the subject. In this context, the broader intellectual environment portrayed at Wittenburg is noteworthy as it underscores the magnitude of Faustus's newly acquired cosmographical knowledge.

The high regard in which Faustus is held upon his return to this environment of respectable scholarship evidences his ascension to even greater heights of traditional academic institutional esteem than previously thought. As ciphers for a wider intellectual community, the promotion of Faustus in matters of 'astrology' to the status of teacher and pedagogue reveals powerfully the impact of the practical cosmography on his scholarly eminence. This authority is such that it extends beyond the bounds of the university: 'Now his fame spread forth in every land' (4. Chorus. 12). Pre-eminence in cosmography, previously elusive under the constrictions of scholarly disputation, has been achieved through practical exploration.

Marlowe's delineation of Faustus, the cosmographer, has been seen as imbued with a 'tension between heroic ambition and ironic failure'.[97] It is evident that the protagonist's attempts at gaining cosmographical knowledge purely through disputation with his interlocutor Mephistopheles are unsuccessful. Faustus is a failed cosmographer, fruitlessly seeking ultimate truths through a discursive interrogation of the cosmos's structure. But in his practical cosmography, Faustus is a heroic figure, his empirical endeavours substantiating the bookish learning of his previous self and serving as an authority on all matters cosmographical to both his university colleagues and others throughout Europe. In depicting Faustus in such a fashion, Marlowe displays his sensitivity to the prevailing conception of the successful cosmographer as both a scholar of bookish learning and an active explorer whose experience informs his knowledge. In a fashion which strikingly prefigures Bacon's theory/practice of educational symbiosis, it is only through a combination of both that Faustus, the eminent cosmographer, is fully realised. As such, Marlowe dramatises another prevailing notion of the geographic consciousness of the age, presenting powerfully to the audience the wide-ranging potentials of cartography.

Notes

1. William Hazlitt, 'Lectures on the Dramatic Literature in the Age of Elizabeth', *Marlowe: The Critical Heritage, 1588–1896*, ed. Millar McClure (London: Routledge 1979), pp. 77–78.
2. See Marcie Bianco, 'To Sodomize a Nation: *Edward II*, Ireland, and the Threat of Penetration', *EMLS*, 11 (1997), pp. 11.1–21. Bianco identifies Gaveston, one of the play's main characters, 'as the metonymic embodiment of Ireland [...] [who] comes to figure as the nodal point where Ireland and sodomy intersect in *Edward II*' (para. 1). For another analysis of Ireland's role in Marlowe's play, see Stephen O'Neill, *Staging Ireland: Representations in Shakespeare and Renaissance Drama* (Dublin: Four Courts Press, 2007), pp. 90–103.
3. Ben Jonson quoted in Charles Nicholl, 'Marlowe, Christopher (bap. 1564, d. 1593)', *ODNB* (Oxford: Oxford University Press, 2004), http://www.oxforddnb.com/view/article/18079 (accessed 21st August 2018).
4. Michael Neill, '"Mulattos", "Blacks" and "Indian Moors": *Othello* and Early Modern Constructions of Human Difference', *Shakespeare Quarterly*, 49:4 (1998), p. 365. *Hungarians, Germans, Italians (Faustus), Scythians, Babylonians (Tamburlaine), English, Scottish, Welsh (Edward II), Greeks, Turks, Carthaginians, Jews (The Jew of Malta) and many others all intermingle – sometimes harmoniously, though often discordantly – in the playwright's works.*
5. Richard Willes in *History of Travayl* by Richard Eden (London: Richarde Jugge, 1577), sigs. I1^r–2^v.

6. Irving Ribner, 'Marlowe's "Tragicke Glasse"', *Essays on Shakespeare and Elizabethan Drama in Honor of Hardin Craig*, ed. Richard Hosley (Columbia: University of Missouri Press, 1962), p. 91.

7. Jonathan M. Smith, 'State Formation, Geography, and a Gentleman's Education', *Geographical Review*, 86:1 (1996), p. 97.

8. For an extensive account of the staging history of Marlowe's play, see the entry for *Tamburlaine* in volume two of *British Drama, 1533–1642: A Catalogue*, ed. Martin Wiggins (Oxford: Oxford University Press, 2012), pp. 375–379.

9. David McInnis, *Mind-Travelling and Voyage Drama in Early Modern England* (Basingstoke: Palgrave Macmillan, 2012), p. 51.

10. McInnis, *Mind-Travelling*, p. 7.

11. This short summary of the real and allusive geographies of the play indicates its variety: beginning in Persia, the action passes through Turkey, Parthia, Mesopotamia, Asia (from Tartary to Russia to Colchis), India, the East Indies, Egypt, Africa (from Morocco to Tripoli to the Barbary Coast), Constantinople, Greece, Europe (from Albania to Venice to Gibraltar to Britain), the Mediterranean, Mexico, the West Indies and the Antarctic Pole.

12. Mark Thornton *Burnett*, 'Tamburlaine: An Elizabethan Vagabond', *Studies in Philology*, 84:3 (1987), p. 321.

13. William Cuningham, *The Cosmographical Glasse conteinyng the pleasant principles of cosmographie, geographie, hydrographie, or nauigation* (London: John Day, 1559), p. 120. As Emrys Jones remarks, 'Marlowe's geographical awareness will strike any reader of the Tamburlaine plays'. See Jones, '"A World of Ground": Terrestrial Space in Marlowe's *Tamburlaine* Plays', *The Yearbook of English Studies*, 38:1–2 (2008), p. 168.

14. Jones, '"A World of Ground"', p. 168.

15. Ethel Seaton, 'Marlowe's Map', *Essays and Studies*, 10 (1924), pp. 13–35 and 'Fresh Sources for Marlowe', *The Review of English Studies*, 5:20 (1929), pp. 385–401.

16. Jones, '"A World of Ground"', p. 168.

17. Stephen Greenblatt, *Renaissance Self-fashioning: From More to Shakespeare* (Chicago: Chicago University Press, 1980), pp. 200–201.

18. Greenblatt, *Renaissance Self-fashioning*, p. 201.

19. John Gillies, 'Marlowe, The Timur Myth and the Motives of Geography', *Playing the Globe: Genre and Geography in English Renaissance Drama*, eds. John Gillies and Virginia Mason Vaughan (Madison, New Jersey: Fairleigh Dickinson University Press, 1998), p. 226.

20. Garrett Sullivan, 'Geography and Identity in Marlowe', *The Cambridge Companion to Christopher Marlowe*, ed. Patrick Cheney (Cambridge: Cambridge University Press, 2004), p. 232.

21. Sullivan, 'Geography and Identity in Marlowe', p. 232.

22. Lisa Hopkins, *Christopher Marlowe, Renaissance Dramatist* (Edinburgh: Edinburgh University Press, 2008), p. 98.

23. Hopkins, *Christopher Marlowe, Renaissance Dramatist*, p. 98.

24. Hopkins, *Christopher Marlowe, Renaissance Dramatist*, p. 100.

25. A contrasting perspective on such themes is presented in Gregory Woods, 'Body, Costume, and Desire in Christopher Marlowe', *Journal of Homosexuality*, 23:1–2 (1992), pp. 69–84. Woods emphasises the bodily landscape, foregrounding what he terms Marlowe's 'erotic geography' by analysing the Renaissance analogue of mapped space and anatomised body that I have explored in the previous chapter on Speed and Drayton: Gaveston in *Edward II* 'marks a crucial spot in Marlowe's erotic geography, equidistant between devilry and innocence, pastoral and decadence' (p. 82).

26. Jonathan Sumption, *Age of Pilgrimage: The Medieval Journey to God* (Mahwah, New Jersey: Paulist Press, 2003), p. 214.
27. Richard Hardin, 'Marlowe Thinking Globally', *Christopher Marlowe the Craftsman: Lives, Stage, and Page*, eds. Sarah K. Scott and M.L. Stapleton (Surrey: Ashgate, 2010), pp. 23–32.
28. *Smith,* 'State Formation, Geography, and a Gentleman's Education', p. 94.
29. Thomas Blundeville, *His Exercises* (London: John Windet, 1594), p. 134.
30. Lesley B. Cormack, 'Maps as Educational Tools in the Renaissance', *The History of Cartography, Volume Three, Part 1: Cartography in the European Renaissance*, ed. David Woodward (Chicago: Chicago University Press, 2007), p. 622.
31. Nicholl, *ODNB*. For a more detailed description of Gresshop, see Constance Brown Kuriyama, *Christopher Marlowe: A Renaissance Life* (Ithaca: Cornell University Press, 2010), pp. 25–28. Kuriyama describes the extent of Gresshop's bibliophilia: 'When we add up the sums at which his books and clothes were valued, we discover that Gresshop spent approximately twice as much on books as on clothes, at a time when clothes were quite expensive' (p. 27).
32. Vivien Thomas and William Tydeman, *Christopher Marlowe: The Plays and Their Sources* (London: Routledge, 1994), p. 5.
33. Thomas and Tydeman, *Christopher Marlowe*, pp. 5–12. For an examination of the parallels between the popular morality plays staged in England in Marlowe's lifetime and the playwright's writings, see Jeff Dailey, 'Christian Underscoring in *Tamburlaine the Great, Part II*', *The Journal of Religion and Theatre*, 4:2 (1995), pp. 146–159.
34. Matthew Adam McLean's *The Cosmographia of Sebastian Münster: Describing the World in the Reformation* (Surrey: Ashgate, 2007) represents an informative discussion of the role of *Cosmographie* in the growth of sixteenth-century European cosmography. For an overview of Münster's life and work, see Margaret T. Hodgen, 'Sebastian Münster (1489–1552): A Sixteenth-Century Ethnographer', *Osiris*, 11 (1954), pp. 504–529. Columbia University's online database of Münster's work represents an introduction to his writings (http://www.columbia.edu/itc/mealac/pritchett/00generallinks/munster/munster.html (accessed 14th February 2014).
35. Benjamin Weiss, 'The *Geography* in Print: 1475–1530', *Ptolemy's 'Geography' in the Renaissance*, eds. Zur Shalev and Charles Burnett (London: Warburg Institute Colloquia 2011), p. 110.
36. For a discussion of *Cosmographie*'s 'encyclopaedism', see Elizabeth Eisenstein, *The Printing Press as an Agent of Change*, Volume 1 (Cambridge: Cambridge University Press, 1980), p. 109.
37. Richard Eden, *A briefe collection and compendious extract of the straunge and memorable things, gathered oute of the cosmographye of Sebastian Munster* (London: Thomas Marshe, 1572), sig. A2^r.
38. Lesley B. Cormack, *Charting an Empire: Geography at the English Universities 1580–1620* (Chicago: University of Chicago Press, 1997), p. 107, n. 55.
39. For an outline of the hybridity of this text, see Cormack, *Charting an Empire*, p. 129.
40. This facet of Marlowe's life has been keenly debated. For an overview, see Nicholl, *ODNB*. See also Austin K. Gray, 'Some Observations on Christopher Marlowe, Government Agent', *PMLA*, 43:3 (1928), pp. 682–700; and Alan Haynes, *Invisible Power: The Elizabethan Secret Services, 1570–1603* (New York: St. Martin's Press, 1992), pp. 99–100.
41. Lyne's map has been singled out by P.D.A. Harvey as an 'unusually clear example' of Tudor cartography's combination of bird's-eye perspective and

consistent scale (Harvey, *Maps in Tudor England* (Chicago: Chicago University Press, 1993), p. 17).

42. Jeffrey John Speed, *Tudor Townscapes: The Town Plans from John Speed's 'Theatre of the Empire of Great Britaine 1610'* (Buckingham: Map Collector Publications Ltd, 2000), p. 32.

43. Mark Curtis, *Oxford and Cambridge in Transition: 1558–1642* (Oxford: Clarendon Press, 1959), p. 130.

44. Curtis, *Oxford and Cambridge in Transition*, p. 130.

45. Cormack, *Charting an Empire*, p. 307.

46. Vivien Thomas and William Tydeman call attention to the importance of Münster's textbook in the source material for Marlowe's plays: 'the heterogeneous materials laid under debt in the second part of Tamburlaine must include Münster's Cosmographie' (Thomas and Tydeman, *Christopher Marlowe: The Plays and Their Sources* (1994) (London: Routledge, 1994), p. 80). For broader discussions of the status of geography in the English universities in the Renaissance and the educational teaching of humanism, see J.N.L. Baker, 'Academic Geography in the Seventeenth and Eighteenth Centuries', *Scottish Geographical Magazine*, 51:3 (1935), pp. 129–144. David Riggs highlights the more practical political and vocational consequences of geography as a subject in university. See Riggs, *The World of Christopher Marlowe* (Basingstoke: Macmillan, 2004), p. 159.

47. J.C.T. Oates, *Cambridge University Library: A History* (Cambridge: Cambridge University Press, 1986), pp. 115–116.

48. Oates, *Cambridge University Library*, pp. 122–123.

49. Camden's 'squat octavo volume', a combination of chorography, cartography, history and mythography, has been identified as a seminal text in the development of 'English multi-disciplinary historical writing' (Wyman H. Herendeen, 'Camden, William (1551–1623)', *ODNB* (Oxford: Oxford University Press, 2004), http://www.oxforddnb.com/view/article/4431 (accessed 21st February 2019)).

50. Cormack, *Charting an Empire*, pp. 55–56. As Catherine Delano Smith's research has shown, private map ownership was a growing feature of life at university. See Delano Smith, 'Map Ownership in Sixteenth-Century Cambridge: The Evidence of Probate Inventories', *Imago Mundi*, 47 (1995), pp. 67–93.

51. Cormack, *Charting an Empire*, p. 227.

52. Cormack, *Charting an Empire*, pp. 55–56.

53. Cormack, *Charting an Empire*, p. 40.

54. Michael G. Moran contextualises Cuningham's guide within the development of sixteenth-century English land surveying and mapping. See Moran, *Inventing Virginia: Sir Walter Raleigh and the Rhetoric of Colonization, 1584–1590* (New York: Peter Lang Publishing, 2007), pp. 181–202.

55. See, for example, *De Planetarium* (1550) and *An Astronomical Treatise* (1553). For an extensive examination of Dee's geographical writing and activities, see E.G.R. Taylor's study *Tudor Geography 1485–1583* (London: Methuen and Company, 1930), pp. 75–139.

56. Nicholls, an explorer and merchant by profession, also engaged in geographical writing, and his travel narratives and chorographical descriptions of places such as Peru, China and the West Indies were incorporated into Richard Hakluyt's popular *Principall Navigations* published towards the end of the sixteenth century (R.C.D. Baldwin, 'Nicholls, Thomas (1532–1601)', *ODNB* (Oxford: Oxford University Press, 2004), online edn. Jan 2008 http://www.oxforddnb.com/view/article/20124 (accessed 21st February 2014)). Hood

was a close associate of the famed English globemaker Emery Molyneux, and published practical guides to the use of cartographic instruments including *The Use of the Celestial Globe in Plano, Set Foorth in Two Hemispheres* (London: John WIndet, 1590) 'largely intended to help the student astronomer to recognize the stars and their constellations' (H.K. Higton, 'Hood, Thomas (*bap.* 1556, *d.* 1620)', *ODNB* (Oxford: Oxford University Press, 2004), online edn. Jan 2008 http://www.oxforddnb.com/view/article/13680 (accessed 21st February 2014)) and also *The Use of Both the Globes Celestiall and Terrestriall* (London: Thomas Dawson, 1592).

57. Two other key instances of map-reading in early modern drama which are not explored in this thesis occur in plays by Shakespeare – *Henry VI Part I* (3.1.3–104) and *King Lear* (1.1.25–30). For readings of the map's role in these respective plays, see Anthony Roche, 'A Bit off the Map: Brian Friel's *Translations* and Shakespeare's *Henry IV*', *Literaray Interrelations: Ireland, England and the World*, Volume 2: Comparison and Impact, eds. Wolfgang Zach and Heinz Kosok (Tubingen: Gunter Narr Verlag, 1987), pp. 139–148, Matthew Greenfield, '*1 Henry IV*: Metatheatrical Britain', *British Identities and English Renaissance* Literature, eds. David J. Baker and Willy Maley (Cambridge: Cambridge University Press, 2002), pp. 71–80, F.T. Flahiff, 'Lear's Map', *Cahiers Élisabéthains*, 30 (1986), pp. 17–30, Terence Hawkes, *Meanings by Shakespeare* (London: Routledge, 1992), pp. 121–140 and John Gillies, 'The Scene of Cartography in *King Lear*', *Literature, Mapping, and the Politics of Space in Early Modern Britain*, eds. Andrew Gordon and Bernhard Klein (Cambridge: Cambridge University Press, 2001), pp. 109–137. These cited discussions are only a selection of what is a substantial body of scholarship.

58. All quotations are taken from Christopher Marlowe, *Doctor Faustus and Other Plays*, eds. David Bevington and Eric Rasmussen (Oxford: Oxford University Press, 1995).

59. See, for example, Tom Rutter, *The Cambridge Introduction to Christopher Marlowe* (Cambridge: Cambridge University Press, 2012), pp. 26–36.

60. Robert Burton, *The Anatomy of Melancholy* (Oxford: Henry Cripps, 1621), p. 436.

61. Seaton, 'Marlowe's Map', p. 13. The influence of Seaton's identification of Ortelius's presence within the play on interpretations, and subsequent editors of Marlowe, is indicated by Una Ellis-Fermor's edition of *Tamburlaine the Great: In Two Parts* (London: Methuen, 1930) which includes as a foldout a facsimile of two maps from Ortelius's atlas.

62. Cuningham, *The Cosmographical Glasse*, sig. A5^v.

63. I have included in this count Tamburlaine's citation of circles of latitude, 'the tropic line of Capricorn' (138) and 'Cancer's line' (146), as while they represent more general signifiers of location as opposed to specific cities, countries and seas, they also illustrate the varied nature of the geographic regions cited by the protagonist.

64. Cuningham, *The Cosmographical Glasse*, sig. A5^v.

65. Burton, *Anatomy of Melancholy*, p. 4.

66. Lisa Hopkins, *Christopher Marlowe: A Literary Life* (Basingstoke: Palgrave Macmillan, 2000), p. 45.

67. Hopkins, *Christopher Marlowe*, p. 45.

68. The original published edition of the play, entitled *The Second Part of The Bloody Conquests of Mighty Tamburlaine*, introduces the map speech with the stage directions 'Alarme, Tamb. Goes in, and comes out with al the rest'. This is maintained (with the minor inclusion of the word 'againe' in the phrase 'comes out [againe] with al the rest') in the subsequent sixteenth and

seventeenth-century published editions of the play (1592, 1597, 1605). A standard of more recent editions of Marlowe's play, such a stage direction indicates a crowded theatrical scene, emphasising the fact that Tamburlaine's map-reading is a highly public event.

69. According to Denis Wood, Marlowe's map-reading provides a 'reality that exceeds our vision, our reach, the span of our days, a reality we achieve no other way'. See Wood, *Rethinking the Power of Maps* (New York: Guilford Press, 2012), p. 4.

70. Sara Munson Deats, 'Myth and Metamorphosis in Marlowe's *Edward II*', *Texas Studies in Literature and Language*, 22:3 (1980), p. 305. Marlowe's extensive use of myths and classical sources in both poetry and drama has been well-documented. Eugene M. Waith's study *The Herculean Hero in Marlowe, Chapman, Shakespeare and Dryden* (New York: Columbia University Press, 1962) explores Marlowe's usage of the Herculean prototype, while Christopher Wessman's 'Marlowe's *Edward II* as "Actaeonesque History"', *Connotations*, 9:1 (1999), pp. 1–33, traces parallels between the doomed huntsman of mythology and the tragedian monarch. For other case studies of Marlowe's employment of classical figures, themes and tropes, see also Paul W. Miller, 'A Function of Myth in Marlowe's "Hero and Leander"' *Studies in Philology*, 50:2 (1953), pp. 158–167; and Tom McAlindon, 'Classical Mythology and Christian Tradition in Marlowe's *Doctor Faustus*', *Publications of the Modern Language Association of America*, 81:3 (1966), pp. 214–223, especially pp. 217–223.

71. Marlowe, *Tamburlaine*, part 1, 4.4.78–83.

72. John Gillies, *Shakespeare and the Geography of Difference* (Cambridge: Cambridge University Press, 1994), p. 56.

73. Cormack, *Charting an Empire*, p. 90.

74. Cormack, *Charting an Empire*, pp. 92–93.

75. According to James McDermott, Martin Frobisher, arguably Elizabeth I's most trusted naval general, was 'uncultured [,] semi-literate [and] unversed in the hydrographical sciences'. James McDermott, 'Frobisher, Sir Martin (1535?–1594)', *ODNB*, (Oxford: Oxford University Press, 2004), online edn. January 2008, http://www.oxforddnb.com/view/article/10191, (accessed 23rd December 2020).

76. Walter Raleigh, *The History of the World* (London: William Stansby, 1617), sig. E4[r].

77. Raleigh, *The History of the World*, p. 330.

78. Peter Holland, 'The Dramatic Forms of Journey in English Renaissance Drama', *Travel and Drama in Shakespeare's Time*, eds. Jean-Pierre Maquerlot and Michèle Willems (Cambridge: Cambridge University Press, 2006), p. 161.

79. McInnis, *Mind-Travelling and Voyage Drama in Early Modern England*, p. 66.

80. McInnis, *Mind-Travelling and Voyage Drama in Early Modern England*, p. 66.

81. For the theological underpinnings of this debate, see Francis R. Johnson, 'Marlowe's Astronomy and Renaissance Skepticism', *ELH*, 13:4 (1946), pp. 241–254.

82. David Riggs observes: 'In theory MA-level work on astronomy, geography and cosmography taught aspiring divines to know the Creator through the study of His works'. Riggs, *The World of Christopher Marlowe*, p. 159.

83. Jerry Brotton, *A History of the World in Twelve Maps* (London: Penguin 2012), p. 242.

84. Thomas Blundeville, *M. Blundevile His Exercises Containing Six Treatises* (London: John Windet 1594), p. 134.

85. Petrus Ramus, *The way to geometry Being necessary and usefull, for astronomers*, trans. William Bedwell (London: Thomas Cotes, 1636), sig. A5[v].

86. Dee, *The elements of geometrie of the most auncient philosopher Euclide of Megara* (1570), sig. *1ʳ.
87. Paul H. Kocher, 'The Old Cosmos: A Study in Elizabethan Science and Religion', *The Huntington Library Quarterly*, 15:2 (1952), pp. 105–106.
88. In a notable interpretation of these events, David Kimball Smith argues that Faustus's attention is resolutely earthly and is 'focussed on knowledge of the world: not just the metaphorical "world of profit and delight," but the physical world'. Smith, *The Cartographic Imagination in Early Modern England: Re-Writing the World in Marlowe, Spenser, Raleigh and Marvell* (Surrey: Ashgate, 2008), pp. 147–148.
89. Stephen Orgel, 'Tobacco and Boys: How Queer Was Marlowe?' *GLQ: A Journal of Lesbian and Gay Studies*, 6:4 (2000), p. 569.
90. Cuningham, *The Cosmographical Glasse*, sig. A5ᵛ.
91. Cuningham, *The Cosmographical Glasse*, sig. A4ʳ.
92. While C.L. Barber's essay 'The Form of Faustus' Fortunes Good or Bad', *The Tulane Drama Review*, 8:4 (1964), pp. 92–119, has stressed interpretations of the play as an 'expression of the Reformation [...] [and] a Promethean enterprise, heroic and tragic, an expression of the Renaissance' (p. 92), others have explored the fundamental inadequacies of Faustus's intellectual character. See Leo Kirschbaum, 'Marlowe's *Faustus*: A Reconsideration', *Review of English Studies*, 19:75 (1943), pp. 225–241, which describes Faustus as an 'unstable, foolish worldling' (p. 240), and Phoebe S. Spinrad, 'The Dilettante's Lie in *Doctor Faustus*', *Texas Studies in Literature and Language*, 24:3 (1982), pp. 243–254. Jonathan Dollimore in *Radical Tragedy: Religion, Ideology, and Power in the Drama of Shakespeare and His Contemporaries* (North Carolina: Duke University Press, 2004) offers a more ambivalent reading of the play's narrative, emphasising the irreconcilable ambiguities in the characterisation of Faustus (pp. 109–119).
93. Andrew J. Sofer, 'How to Do Things with Demons: Conjuring Performatives in *Doctor Faustus*', *Theatre Journal*, 61:1 (2009), p. 10.
94. Marlowe, *Doctor Faustus and Other Plays*, p. xiii.
95. Interpreting this passage within the context of the play's broader cosmographic themes, Donald Kimball Smith reminds us of the extent of this journey and the centrality of its importance for Faustus's comprehension of geography and cosmography: 'Although his voyage is a good deal briefer than Raleigh's voyage to Guiana, Faustus discovers and explores not just a New World, but the whole world. He centres much of his desire for knowledge in the cartographic understanding of the world's geography, and in this pursuit, magic becomes a kind of instant travel, a means for the discovery of different lands'. Smith, *The Cartographic Imagination in Early Modern England*, p. 148.
96. Francis Dolores Covella, 'The Choral Nexus in *Doctor Faustus*', *Studies in English Literature, 1500–1900*, 26:2 (1986), p. 210.
97. Troni Y. Grande, *Marlovian Tragedy: The Play of Dilation* (Lewisburg: Bucknell University Press, 1999), p. 181, n. 3.

7 Wenceslaus Hollar's cartographies

Having considered how Marlowe exhibited mapping on the stage, I would now like to consider mapping on the page. In particular, this chapter focuses on the cartographies of the engraver Wenceslaus Hollar and its exhibition of the array of styles, techniques and methods of representation of Renaissance cartography. Surveying his work, it argues that the artist demonstrates the variegated typologies of early modern map-making discussed throughout this study, emblematising a discipline encompassing a rich diversity of representative modes and attendant motivations.

'he tooke a delight in draweing of mapps': Hollar in a cartographic context

Born in Bohemia in 1607, Wenceslaus Hollar moved to England with his patron Thomas Howard, second earl of Arundel, in 1636 following the latter's ambassadorial visit to the court of the emperor Ferdinand II.[1] An accomplished draughtsmen who, according to Howard, 'drawes and eches printes in strong water quickely, and w[i]th a pretty spiritte', Hollar worked initially as a portraitist, documenting women's fashion and displaying his talent as a copyist in reproductions of the works of continental artists such as Marcus Gheerearts, Albrecht Dürer and Leonardo Da Vinci.[2] Hollar's etching career extended from the 1620s to the 1670s and is as wide-ranging as it is extensive, with his work displaying an assortment of styles and subjects.[3]

In keeping with many other early modern figures, however, Hollar was particularly enthused by cartography and frequently employed his aptitude for draughtsmanship in the production of maps. As George Vertue's extensive assembling of Hollar's 'maps, plans, views, and prospects' and Richard Pennington's exhaustive *Descriptive Catalogue of the Etched Work of Wenceslaus Hollar 1607–1677* (1982) indicate, Hollar was responsible for a substantial body of cartographic work.[4] Nonetheless, Hollar's considerable mapping endeavours have received little sustained attention by cartographic historians and scholars of early modern cartography. This is surprising, particularly when we consider the extent of Hollar's

DOI: 10.4324/9781003200376-11

maps – Vertue, in particular, dedicates more pages in his catalogue to Hollar's maps than any other subject. Pennington's catalogue also reveals the variety of this particular section of the author's work, listing under the general collocation 'Geography and Maps' various different generic classifications including (but not limited to) maps, plans and views (of both urban and agrarian landscapes), landscapes 'after famous painters' and seascapes.[5]

Additionally, Hollar's biographical background is marked by an affinity for the cartographic consciousness of the age. Hollar's enthusiasm for mapmaking was in evidence from an early age. Including him in his popular biographical compendium *Brief Lives*, his friend John Aubrey records Hollar's own admission that his interest in maps started early: 'He told me that when he was a Schoole-boy he tooke a delight in draweing of Mapps; which draughts he kept, and they were pretty'.[6] Craig Hartley, referencing Aubrey's short profile, and drawing on Hollar's familial circumstances, has alerted us to the cartographic background from which the etcher emerged, asserting that 'Hollar's earliest predilection was for topography and map-making'. Hartley continues:

> A relative of Hollar [Pavel Aretin, Hollar's uncle by marriage] was a publisher of maps and trained in cartography, while Hollar's father, Jan, who worked in the records office in Prague Castle, accompanied land surveyors when estates in the region had to be mapped.[7]

This familiarity with the administrative function of surveying and mapping would manifest itself later in Hollar's own production of topographical and cartographical drawings for the express purpose of civic governance. For example, his etched bird's eye view plan of London was designed as a survey of the city before the Great Fire for the consumption of local civic official Robert Viner, one-time alderman and Lord Mayor of the City.

Scholars have also discerned in Hollar's continental origins a persistent fascination for mapmaking that predisposed the young draughtsman to employ his considerable skills in the pursuit of cartography.[8] Significantly for his connections to the flourishing discipline of Renaissance cartography, Hollar's training in Frankfurt between 1627 and 1629 was carried out under the tutelage of Mattheus Merian. A copperplate engraver of patrician stock, Merian's chief vocation was cartographically inclined, involving the designing of 'prospects' or townscapes.[9] The townscape, with its delineation of space from a standardised viewpoint (in this case, ground-level panorama views), was one of the many types of cartographically inclined genres that would later populate Hollar's work. The etcher's associations in adulthood reveal a further familiarity with intellectuals who cultivated an enthusiasm for cartography – acquaintances included map-collectors like the diarist John Evelyn, who maintained an interest in map-making and was known to annotate maps.[10] Taken together, Hollar's own personal enthusiasm for cartography, his familial and cultural background, which was occupied professionally in the production and dissemination of maps and his adult associations mark out a figure deeply immersed in the endeavour of the world describing. This immersion would manifest itself in a wide-ranging cartographic output, one sensitive to the generic characteristics of different types of maps and, concomitantly, exhibiting the varying discourses in which maps could be used.

Hollar's professional cartography in a biographical context

Numbering over 600 individual engravings, Hollar's maps are as varied as they are vast. His cartographic output, in many ways, encapsulates the diversification in European cartography from the 1500s onwards: 'Nor was [it] just a revolution in the acceptance and use of maps,' states P.D.A. Harvey of the early modern period, 'it was equally a revolution in the kind of map that was produced'.[11] This 'cartographic revolution' benefited from advances in technology, wider distribution of geographical texts and also improvements in cartographic and surveying processes. Additionally, it accrued from a coming together of different ideas as to what constituted a map. Increasingly in the early modern period, the idea of the map underwent a transformation, and according to Harvey, this development was engendered by a standardisation of representational parameters, and in particular, the increasingly prevalent use of scalar measurement. Harvey continues:

> Looking still at maps of small areas – a house, a field, a town, a tract of countryside, an entire country – it is no exaggeration to say that the map as we understand it was effectively an invention of the sixteenth century. Earlier maps might be itinerary maps, showing the succession of points along one or more routes but not what lay between the routes or how to get from one route to another. Or they might be picture maps, showing features above ground level pictorially as though seen from a height on bird's-eye view, with more or less attention to detail and varied degrees of artistry – sometimes very little. Or, if they showed only the outlines of buildings or fields, these would be drawn roughly to shape, sometimes with measurements written in, but never consistently to scale.[12]

The homogenising effect of the employment of scale on cartography as a whole is singularly transformative if we consider, as Harvey proposes ('Looking still at maps of small areas', 'the map as we understand it'), cartography and its underlying motivations from a singular perspective. Such a viewpoint would stipulate mapping's innate accuracy, realistic reproduction of geographical area and socio-cultural and politically non-partisan production. This perception of cartography as, in the words of the modern poet Eavan Boland 'the masterful, the apt rendering | of the spherical as flat' has persevered.[13]

Harvey's enumeration of the prevailing generic variation of mapping prior to the sixteenth century – including itinerary maps, bird's-eye views and panoramic vistas – also alludes to the diversity of mapping techniques, modes, conventions and methodologies accessible to early modern mappers. Homogeneity was in many ways an alien concept to Renaissance cartography, with the development of the world describing in the early modern period being characterised by both evolution and revolution. Early modern maps may have increasingly adopted a uniform representational mode and scaled adumbration of landscape from a standardised viewpoint. However, more recognisably pre-modern mapping modes such as allegorised representations of the landscape (as we have seen in Drayton's frontispiece) and depictions which deviated from the increasingly prevalent and now-standard north-up orientation were also widely employed. According to Rhonda Lemke Sanford, the incorporation of various mapping abstractions such as 'the cognitive or mental map', 'ceremonial maps' and 'figurative maps' is representative of the varied ways in which 'early modern folks were developing the ability to think cartographically'.[14]

As a producer of maps, Hollar demonstrated a striking adroitness in the employment of established mapping modes for particular purposes. Hollar's cartographic work embodies the heterogeneous character of seventeenth-century mapping. His work exhibits cartography's flexibility as a representative mode enmeshed in a wide range of discourses: while

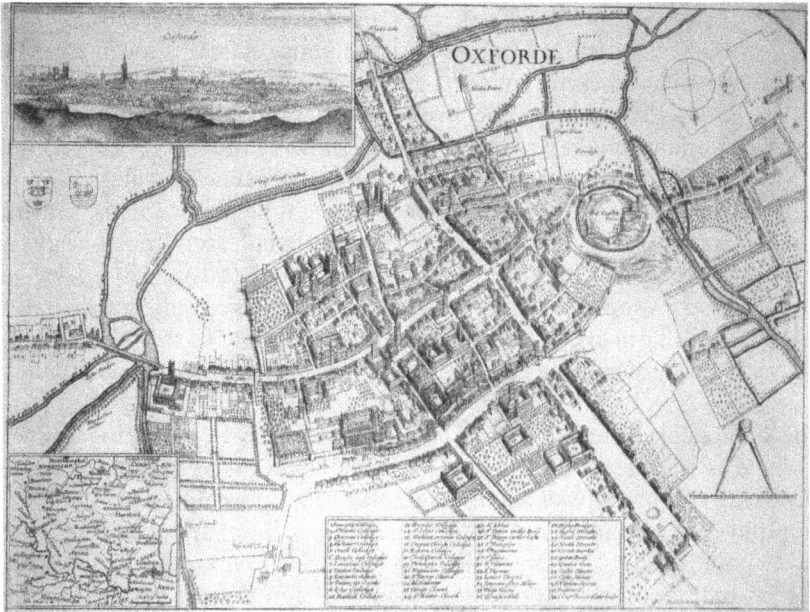

'Oxforde' engraved by Wenceslaus Hollar (1643). (Reproduced by permission of The Thomas Fisher Rare Book Library, Toronto)

the engraver produced many conventional maps, he was also fluent in a whole host of topographical genres, urban and rural, landscapes and seascapes and adduced mapping methodologies for a wide variety of purposes including aggrandisement, decoration, propaganda and political expression.

Creating varied cartographies

Hollar was capable of creating maps to scale, producing highly conventional projections. In a 1643 town plan of Oxford, the influence of antecedent cartographic conventions is manifest: containing a carefully scaled and detailed layout from a bird's-eye view, the representation of 'Oxforde' incorporates an isometric delineation of buildings, an attached list of numbered annotations referring to streets, landmarks of interest such as churches and a macrocosmic delineation of the city's location in relation to London and the wider area. Notably, the town is shown from various angles, while the map incorporates in the top left-hand corner a 'view' of the city or a landscape panorama. In constructing such a geographic representation, Hollar demonstrates his own skill in producing town views.[15]

Furthermore, in presenting a divergent cartographic typology alongside the standard bird's eye view, the town view also attests to Hollar's awareness of the commercial demands of cartography. According to Gavin Hollis, as whole atlases were often prohibitively expensive, the single sheet map was by far the most affordable and, therefore, widespread cartographic document in early seventeenth-century England.[16] Therefore, published as a single relatively inexpensive sheet, Hollar's survey of Oxford stands as an affordable cartography that addresses one of the most popular attractions of the map in the early modern imagination: the desire for multiplicity.[17] As we have seen, one of the attractions of maps to early modern intellectuals was its apparent capacity to encapsulate and represent in a single glance an otherwise-imperceptible range of topographies. Accordingly, the variety in evidence in the Oxford map, most especially its differing cartographic viewpoints – bird's eye, isometric, town and country – highlights Hollar's responsiveness to the desire among contemporary map-readers for diversity in their cartographies.

From a demographic perspective, Hollar also exhibits his receptiveness to mapping's role as a catalyst and medium for the projection of localised identities. The focused map of Oxford replicates the prevailing English cartographic genre of the detailed town or shire plan. Pioneered in the work of Saxton, this particular mode carried with it a subtext of identity formation and consolidation. Such local maps were imbued with an ordering impulse geared towards the stabilisation of the topography and, by implication, its identity.[18] In his replication of this demarcated, town-centred cartographic gaze, Hollar presents an ordered image of the town. Aggregated by a rounded, comprehensive representation of the locale from various perspectives – both the air and the ground – Hollar's Oxford map replicates the trend in early modern English cartography towards the projection of particularised, demarcated topographies and participates in the broader trend towards the 'instantiation'.

The assemblage of cartographies entitled *Quartermaster's Maps of England and Wales*, published a year after the Oxford map, exhibits another facet of the etcher's nuanced cartographic sensibility. Emblematic of this collection is 'The mappe of Shropshire, Cheshire, Staffordshire, Worcestershire, Herefordshire, Brecknokshire, Caermarthinshire, Cardiganshire, Radnorshire, Montgomeryshire, Merionidhshire, Denbighshire, Flyntshire, Carnarvan, the Ile of Anglesey, & parte of Pembrokeshire'. In this map, Hollar establishes an almost total focus on locational and toponymic specificity. He demonstrates a sensitivity to a particular trend of Renaissance English cartography which gravitated towards the projection of unadorned and, as a consequence, professedly unprejudiced detailing of the landscape. Excising the excesses of the illustrative cartographer, this mode sought to minimise the perception of an intermediary mediation between land and map-reader. As David Harvey explains, with the increasing prominence of 'realistic' maps, a substantial section of geographic science eschewed the aesthetics of decoration in favour of an economic and politically orientated purposefulness:

> [T]he Renaissance tradition of geography as everything understood in terms of space, of *Cosmos*, got squeezed out. [Geography] was forced to buckle down, administer empire, map and plan land uses and territorial rights, and gather and analyse useful data for purposes of business and state administration.[19]

Such a tendency is manifest in Hollar's 1644 surveys, representations which concentrate insistently on one aspect of 'useful data', namely the location and appellations of certain locations. Underlining the utilitarian objectives of the map is the sparseness of superfluous or illustrative adornment.

Hollar's eschewing of such para-textual apparatus as pictographs, heraldic insignia or pictorial semiology is notable in the context of his cartographic scope. As many scholars have observed, ornamental accoutrements form an important element in the representational dynamic of many early modern maps. For example, Paige Newmark has emphasised Christopher Saxton's use of ornamentation and, in particular, his frequent representations of Elizabeth I. In making the monarch a recurrent pictorial trope in his text, Saxton foregrounded her hegemonic status as sole owner of the land: 'Nearly every county promoted her royal hegemony,' asserts Newmark, 'for example if we look at Cornwall, her Royal Coat of arms is both large and prominent, and therefore a strong assertion that the land is hers and hers alone'.[20] Similarly, Jerry Brotton's discussion of European attempts to map the Cape of Good Hope gestures towards the significance of ethnographic representations of native tribes in vistas added to cartographic representations of Southern Africa.[21]

Eschewing such apparatus and distinguishing his work from earlier conventions, Hollar's maps propose a politically impartial objective, one focused exclusively on the recording of places. However, as a paratextual supplement to the *Quartermaster's Maps* implies, the collection's seeming neutrality is destabilised by other non-cartographic elements. Advertised as '[T]he kingdome of England & principality of Wales exactly described whith every sheere & the small townes in every one of them in six mappes' its intended purpose is stated on the title page: 'Useful for all Commanders for quartering of Souldiers and Portable for Every Man's Pocket that would be informed where the Armies be'.[22] The militaristic purpose of Hollar's engraving is evident – the map will be 'Useful for all Commanders' in the quartering of their martial forces and 'Portable for Every Man's Pocket' so that they might be 'informed where the Armies be'. Thus, Hollar's *Quartermaster* cartographies draw our attention to the military usage of cartography in the early modern period and the manner in which paratextual addenda such as title can function in similar ways to much more visual elements such as aestheticised embellishment in the internal dynamic of the map.

As well as military maps and localised town surveys tailored to the desired variety of cartographic representation among his contemporaries, Hollar was also adept at producing predominantly decorative cartographies. Evidencing his sensitivity to the visual pleasure of the map for the early modern cartographic enthusiast, such examples highlight the engraver's alertness to the popular enthusiasm for the aesthetic qualities of the map-maker's art. In 1656, a map by Hollar of the Isle of Man was included in William Smith's *Vale-Royall* (1656). As a more conspicuously ornamental delineation, especially when compared with the *Quartermaster Maps*, it reveals Hollar's awareness of the aesthetic element of the map enthusiasts in the 1650s enjoyed from the experience of map-reading. Containing a dedicatory inscription to Thomas Fairfax, this survey draws on the multi-disciplinary character of early modern geography by presenting an admixture of bird's eye cartographic projection and landscape drawings or topographical scenes. Cartographic enthusiasm from the mid-sixteenth century onwards delighted not only in the accuracy of maps and their functionality but also its willingness to encompass other, seemingly distinct disciplines such as ethnography, antiquarianism, numismatics and even mythology into its delineation. As one of the introductory commendatory preambles to Speed's *Theatre* observes, cartography was conceived as one component of a wider schema of representation:

> This worke therefore being now brought vpon the publike stage and view of the world, may in my poore censure giue satisfaction to the learned Reader, and abide the touch of the malignant opposer; which being effected without hope of gaine or vaine ostentation with so

great care both of body and mind, is the more precious, for, difficilia quae pulchra ['Things that are excellent are difficult']. Wherein Gods power is made knowne in this his weake, but worthy instrument: and the glory of Great Britaine made more famous to the world as well in the Geographicall demension of the Lands situation as in the historicall relations of her most famous monarchs and glorious actions; no Kingdome hitherto so particularly described, nor nations History by true record more faithfully penned.[23]

Here, the necessity of narrative history or 'historicall relations' alongside the 'Geographicall demension' of maps is accentuated, while the presence of historical narrative accounts in Speed's volume is also commended.

In Hollar's projection, eight outlooks surround the central representation of the island, each portraying prominent geographical features, coastal formations and famous ruins. Addressing the map-reader's appreciation of various topographies, the depiction recalls the broad scope of earlier geographically inflected publications. Hollar's dexterity in employing this form in his cartographic projection is especially highlighted by Monique Pelletier's summary of the intricate internal dynamic of the 'perspective view' as a means of urban representation. In particular, Pelletier draws attention the overall functioning of this representative mode in the portrayal of a topographical panorama:

> The city perspective view was based on observation and represented the city in its entirety, giving it shape, the layout of the streets, as well as the appearance and height of the buildings. In addition to those aspects provided by a panoramic view of the city, a perspective view gave a sense of depth and showed the city's streets passing between the rows of houses. The author of a perspective view, which was an intermediary between the panorama and the orthogonal or geometric plan, created the illusion of a single viewpoint, giving a comprehensive view of the city, but the reality was more complex. The author constructed, with ingenuity, various portions from different viewpoints and assembled the constituent elements.[24]

Drawing together geological, historical and archaeological sites of interest and amalgamating their diversities into a single delineation of the island, Hollar constitutes a singular yet multi-faceted image of the landscape. He adopts the role of city shaper, projecting an urban environment at once in planar view and teeming with socio-cultural character.

Moreover, the diversity in the character of the scenes reiterates the multi-disciplinary focus of mapping texts such as Speed's *Theatre*. Ralph Hyde draws our attention to the prevailing Renaissance interest in the view, a trend which stemmed to a large degree from their sanitised delineations of urban and rural landscapes. 'In prospects of British towns one finds no slums, no open sewers, no evidence of crime or poverty,' writes Hyde of the ulterior factors motivating passion for the genre, 'fine architectural landmarks predominate. Town prospects are almost always gilded'.[25] The central projection itself is, importantly for our survey of Hollar's cartographic consciousness, not an original Hollar production but instead a copy of a previous map.[26] However, Hollar's pointed inclusion of various pictographs of sites of topographical interest on the island alongside his reproduction of an established projection demonstrates his awareness of the early modern map-reader's enthusiasm for a diverse outlook in their cartographic consumption. In gilding the map with sanitised and largely depopulated topographical scenes, Hollar shows his alertness to popular modes of cartographic representation such as Speed's *Theatre,* which presents to the reader a demographic and topographic representation of the land.

Alongside prospects depicting local landmarks such as Castle Rushen and geological features such as the Calf of Man, the Isle of Man map also incorporates a textual chorography, highlighting its hybridised cartographic mode in the vein of Drayton's *Poly-Olbion.* In the bottom left-hand corner, below an inscription of the island's heraldic symbol, Hollar includes a brief description of the island:

> Man by Caesar called Mona by Pliny Monabia by Ptolom [Ptolemy] Monogda, and by Gildas Luboma is an Island seated in the Ocean betwixt England, Scotland and Ireland, it formerly bare the name of a kingdom & hath bene populous & well inhabited very plentifull of cattell Foule and Fish [.] [I]t is now devided into 17 Parishes, many Villages, & defended by two Castells.

This feature, in particular, highlights the alignment of Hollar's production with cartographic traditions. The adjacent positioning of the textual overview of history, geography, genealogy and agriculture to a cartographic delineation replicates the format of Speed's *Theatre,* which combined the perambulatory mode of predecessors such as William Lambarde with extensively surveyed and intricately drawn maps. Hollar's map offers an array of delights to satisfy the cartographic enthusiast – Fairfax could view the island as a whole in a direct bird's eye cartographic projection. He could also take a trip vicariously through its notable prospects, both natural and man-made. Furthermore, he could read about its history and environment in a manner similar to the enthusiastic consumers of Speed's all-encompassing atlas earlier in the century.

Alongside the Isle of Man's heterogeneous projection, Hollar's townscapes serve to highlight his awareness of mathematical principles of geography. Reconciling contradictory geometrical attributes – isometric representation of urban dwellings and a scaled map – these engravings demonstrate another aspect of his cartographic dexterity. As Simon Turner points out, Hollar's knowledge and employment of prevailing methodologies of representation for urban landscapes were substantial: 'Hollar practised all the traditional methods of delineating cities, namely panoramas, prospects or long-views, and ground-plots or bird's-eye views'.[27] Illustrative of this prolific movement between styles and methodologies is Hollar's survey of the Strand, Covent Garden, Lincoln's Inn, Holborn and St Giles-in-the-Field, part of a larger project intended to map 'London and Westminster &c' in the years following the Restoration of 1660. A combination of isometric perspective and meticulously scaled plan, this small section suggests Hollar's full map accorded with the fastidiousness outlined by the initial description of the project to potential investors:

> Propositions Concerning the Map of London and Westminster &c: which is in hand by Wentsel Hollar. This Map is to contain 10 Foot in bredth, and 5 Foot upward wherein shall be expressed not only the Streets, Lanes, Alleyes &c: proportionately measured; but also the Buildings (especially of the principall Houses, Churches, Courts, Halls, etc) as much resembling the likeness of them, as the Convenienceof the roome will permit.[28]

Here, as with the Quartermaster maps, Hollar's self-styling as a surveyor of exactitude is in evidence. 'Correct' replication and scrupulously scaled dimensions are foregrounded, while the amalgamation of plot and the isometric view is indicative of the engraver's ability to utilise two discrete viewpoints – bird's eye and isometric – and their oppositional dimensions to create a more complete picture of his subject. Thus, while the plot affords a scaled view with an accurate street plan 'proportionately measured', the employment of a town view perspective facilitates an adumbration of 'the likeness' of the buildings or 'principall Houses, Churches, Courts, Halls, etc' represented in the map.

Mixing genres: Artistic cartographies

Hollar was also accustomed to the symbolic deployment of cartography in the practice of portraiture in sixteenth and seventeenth-century art, an artistic flourish particularly evident in the aforementioned Marcus Gheerearts' 1592 likeness of Elizabeth I. This can be seen most spectacularly in Hollar's depiction of the Parliamentarian general Robert Devereux, 3rd Earl of Essex (1643). In an echo of the use of cartography as a figurative trope in Gheerearts' Ditchley portrait, Devereux is set against a map of a section of England. Littered with battles won, sieges lifted, surrenders provoked and ambuscades enacted, the portrait employs cartographic representation to map out the successes of the subject, augmenting the triumphant impression established by the armoured general on horseback in the foreground. Analysing the significance of the map in the Ditchley portrait, Albert C. Labriola has set out its function as a visual connection between state power, landscape and the puissant female body of the monarch, alerting us to the symbolic possibilities of cartography transposed onto portraiture.[29] In a similar fashion, the map of Hollar's Devereux evokes the portrayed figure's military status and success, projecting an image both of the human body and also evoking its militaristic character and status.

Finally, we can also discern in Hollar's cartographies an awareness of early modern mapping's susceptibility to non-realist modes of representation. This is evident, especially in his contribution to overtly political texts. For example, a map of Britain by Hollar is included in a Civil War broadside entitled 'A Map and Views of the State of England', which seeks to present 'a comparison of the English and Bohemian Civil Wars with a map of England and view of Prague with scenes of battles and the Battle of the White Mountain of 1620'. The impact of the illustrated element broadside was particularly potent because of its widespread accessibility.[30] Illiteracy proved no disqualification as it might in other mediums: assuming the presence of a ready interlocutor, ease of access was aided by ease of understanding. Such features were characteristic of popular graphic art more generally, including cartography. As Miloš V. Kratochvíl explains, highlighting the significance of illustrated pamphlets to the propaganda of European warfare:

> Just as during the past century and a half the printing of books had made literature accessible to ordinary folk, so now graphic techniques were bringing paintings and drawings of artists within the reach of them, too. They also served to provide exact records for the surveyor, and as a method of book illustration and for the dissemination of information. For example, the whole of the Thirty Years War was accompanied by a pamphlet war, and a large part of the tractates, broadsides, propaganda and polemical pamphlets were illustrated.[31]

Hollar typifies the role played by illustration in broadsides, both as a means of imparting information and also projecting a particular politicised message. This is particularly palpable in the cartographic constituent of the Civil War pamphlet. As with other maps examined, the map of Britain and Ireland lacks the rigorous geographical and toponymic exactitude in evidence in surveys such as the Quartermaster maps. There are no scalar indicators, while the land is devoid of place names.

Yet, while forgoing such specificities of 'correct' cartography, the map's apparent abjuration of verisimilitude nonetheless alerts us to its more symbolic purpose. Hollar's map is not a representation of the landscape but rather a recording of a landscape of battle in the middle of The War of the Three Kingdoms. Thus, a symbolic schema litters the quasi-cartographic landscape: isometric images of troops and figurative tropes – such as the 'Lion Rampant' depicted above the Firth of Forth in northern Scotland – can be discerned, while numerous frigates and men-of-war encircle the

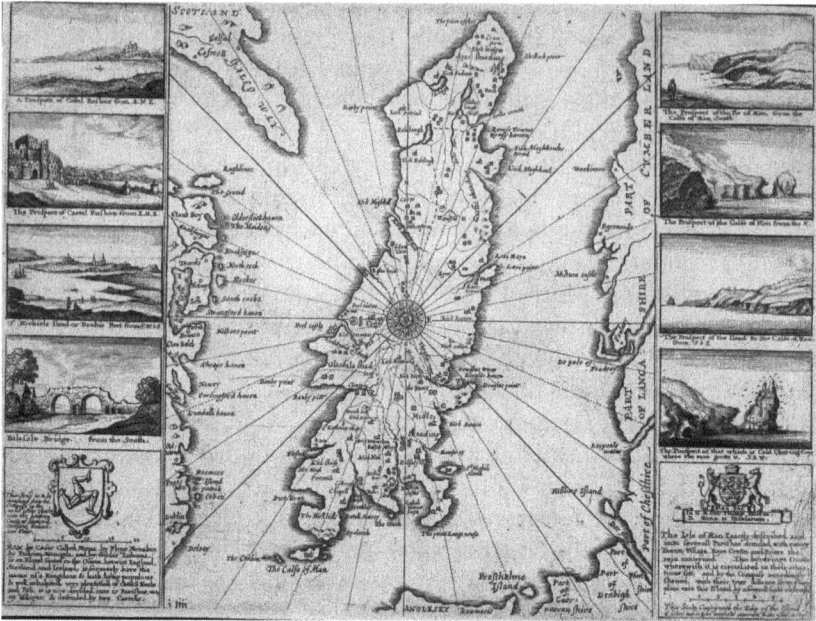

'Isle of Man' engraved by Wenceslaus Hollar. William Smith's Vale-Royall (1656).
(Reproduced by permission of The Thomas Fisher Rare Book Library, Toronto

islands conveying a clear impression of the geography of Britain as a theatre
of conflict. Pointedly, the shape of the land itself is manipulated as cartographies realist imperative is suborned to the map's narrative register. The
theatre is presented as largely Anglo-centric: the extent of the Welsh pen-
insula has been abbreviated, the western seaboard of Scotland and much
of its coastline is omitted, and the island topography of Ireland, though
pullulating with military activity in the seventeenth century, has been contracted. In the minimisation of the margins of the Atlantic archipelago,
Hollar adduces cartography's pliability as a medium of 'real' representation
to evoke an image of a conflict that is predominantly English in character.

Marching alongside soldiers: Hollar's map of Plymouth

Hollar's use of the Civil War as a cartographic subject is also in evidence
in his contribution to a short tract entitled *A True Narration of the Most
Observable Passages, in and at the Late Seige of Plymouth, from the Fifteenth
Day of September 1643, untill the Twenty Fift[h] of December Following.*
While Eudoxus's invocation of the 'plotte' of Ireland in Spenser's *View*
evidences cartography's central role in imperial discourse, Hollar's covering map to this short pamphlet represents a documentable example of J.B.
Harley's description of surveyors as 'march[ing] alongside soldiers' in the
theatre of conflict, 'initially for reconnaissance [...] but also to legitimize

the reality of conquest and empire'.[32] Hollar's 'true map and description of the towne of Plymouth and the fortifications thereof, with the works and approaches of the enemy, at the last seige; A. 1643', endeavours to project a 'true' representation of the siege itself, including military placements, engagements and manoeuvres. Standing as a clear instance of how maps function as tools of warfare, it depicts not only the geographical features of the terrain of the conflict but also the narrative of a conflict as articulated from a particular interested viewpoint.

Following the capture of Bristol in July 1643 by the royalist army, the English Civil War in the southwest entered a new phase of conquest and consolidation. Led by Prince Maurice of the Palatinate, brother to the charismatic royalist commander Rupert, forces loyal to Charles I moved to capture Exeter and then Dartmouth in the late summer. This progress, however, was soon halted at the parliamentarian stronghold of Plymouth, which met its opponents with determined resistance. The subsequent siege lasted more than four months, and was later described by an anonymous author: 'It were endlesse to acquaint you with the servall Skirmishes, that daily past between us, sometimes about our cattel that stragled about our works, at other times, to pass time by bravadoes and ambuscadoes made by our guards to entrappe the enemie'.[33] This account reflects a siege that quickly descended into a protracted daily exchange of attack and counter-attack, artillery fire and cavalry charge. Regularly reinvigorated by ships from the navy under the control of parliamentarians, the town held out until Christmas Day of 1643, when the siege was lifted.[34]

The next year, the 'true narration' was published. As a piece of Civil War history, the textual account itself is not especially remarkable. A steady stream of literature emanated from Plymouth during the early years of the conflict, providing a narrative of the ongoing struggle between parliamentary forces and those loyal to Charles I. In 1642, *Good nevves from Plymouth* was published, 'being a true relation of the death of Sir Ralph Hopton, and many of his commanders, who by treachery fought to surprise the good towne of Plymouth'.[35] Towards the end of that year, a letter from the town to one Captain Joseph Vaughan in London sent at the height of the siege on November 2nd was published in London, and recounts the plight of the town: 'the enemy being masters of the whole County, drawes into their assistance what numbers they please, and hath nowas we heare, brought divers great Guns for battery from Exeter and Dartmouth'.[36] Meanwhile, a year later, a similar text entitled *Joyfull nevves from Plimouth* was published in London, describing itself as 'being an exact relation of a great victory obtained against the Cornish cavaliers'.[37]

Hollar's cartographic contribution, however, marks *A True Narration* out as not just a textual recording of a battle of the Civil War but also a mapped description of events, adducing cartography to express a particular chronicle of the battle. The artist provides a supplementary cartographic delineation of the city and the surrounding area during the latter months of 1643, from Millbrook in the west to Plympton in the east and incorporating Plymouth Sound. What, therefore, does Hollar's contribution to the siege account reveal about his cartographic output? In a relevant remark when considering the interfaces of cartography and conflict, the cultural theorist Edward Said draws our attention to the functioning of geography in wider discourses of self, identity and nationhood. In *Culture and Imperialism*, Edward Said observes:

> I have kept in mind the idea that the earth is in effect one world, in which empty uninhabited spaces virtually do not exist. Just as none of us is outside or beyond geography, none of us is completely free from the struggle over geography. That struggle is complex and interesting because it is not only about soldiers and cannons, but also about ideas, about images and imaginings.[38]

Similarly, Michael Wintle's coherent description of the purpose and intention of cartography and cartographers flags up the subtextual narratives of mapping: 'Maps are not always conscious attempts to present information in a particular light or with an obvious bias (although many are precisely that), [they] will almost invariably reflect unconscious assumptions and assertions to one extent or another'.[39] Ostensibly, Hollar's map offers an example of what Said terms 'images and imaginings'. The map itself, like that of his dedicatory map to Fairfax and the broadside produced earlier in the 1630s, draws together several modes of representation to depict a particular moment of discord in a continuous near-decade-long internecine conflict. The Pennington entry on the map is brief but gives an indication of the variety of the map: 'B. e. v. [bird's-eye view] of the city and the Sound with ships'.[40] It surveys both the social structures surrounding the site of engagement and also the nature of the engagement itself. In accordance with the title – 'Together with an exact map and Description of the Town and Fortifications thereof; with the approaches of the Enemie' – the topographical features of the surrounding landscape of the siege (for example, the coastline) are inscribed with depictions of urban dwellings in the area and also a historical scene of the siege through the delineation of military exercises via sailing frigates and firing ordinance. In doing so, the cartography presents to the reader both an 'image' and an 'imagining' of the siege, displaying simultaneously the geographical expanse and the scene of the Civil War conflict.

The Plymouth map is a noteworthy text in Hollar's cartographic oeuvre because of what it reveals about the fluid and diverse range of its motivations. Furthermore, it underlines the author's attunement to the varied discourses of cartography in the period as set out in this study. Said and Harley, in particular, arguing that maps serve to express a particular subject and draw their motivation from established and interested ideologies, compel us to discern the underlying biases not just of the cartography in itself but also, by extension, the cartographer and socio-political and economic systems underpinning their endeavours.

In this vein, noting the Parliamentarian bias in the narration and Hollar's contribution to it, we might expect the map of the siege to be indicative of the artist's wider loyalties towards the anti-Royalist faction during the Civil War. However, characterising Hollar's personal allegiances as expressly Parliamentarian during the Civil War is problematic. Anthony Griffiths and Gabriela Kesnerová point out that by the early 1640s, Hollar's relationship with his patron Thomas Howard, royal envoy and officer of the court of Charles I, grew increasingly distant:

> [A] larger part of the explanation [for the decrease in commissions] must be that Hollar was never a conventional retainer or employee of the Earl of Arundel. One of his most striking characteristics is the emphasis he lays on his status as a gentleman, and in both his self-portraits he prominently displays his family coat of arms. Such a person might accept patronage, but would never sacrifice his independence.[41]

Suggesting an increasing degree of detachment from Arundel by the start of the Civil War, claims of Hollar's disinterestedness in the politics of the conflict is backed up by Robert J.D. Harding, who notes that Hollar worked independently for a brief period between 1642 and 1644 and for a number of diverse patrons.[42]

Arthur M. Hind, meanwhile, has detected a much more substantial involvement in the conflict. 'In spite of the Civil Wars,' Hind observes, 'Hollar kept busily at work, doing as many as sixty-seven plates in 1643, and forty in 1644'. This industriousness, Hind suggests, was followed by a brief period as a Royalist soldier: 'In the latter year [1644] [Hollar] and several other artists, among whom were Inigo Jones and William Faithorne, took up arms, and served under the Marquis of Winchester at Basing House'.[43] In addition, Hollar spent the start of the conflict in the Royalist stronghold of Bath and in the employ of James Stuart, brother of the Prince of Wales and later James II – a subsequent self-portrait by Hollar emblazoned with the slogan 'Serviteur domestique du Duc de York' suggests an adherence to the Royalists cause.[44]

Consequently, the fact that Hollar affixed his cartography to a Parliamentarian account of a Royalist siege presents a different facet of the artist, one at odds with his earlier evident allegiances to the forces of Charles I. However, what such a contradiction demonstrates is not political schizophrenia but rather the artist's readiness to adapt and adopt differing positions in his professional activities and, most especially, his drawing of maps. Hollar's biographer J. Urzidil reminds us of his artisanal focus during the fraught period of the 1640s and his struggle to maintain a vocational 'haven of peace' amidst the conflict until an eventual departure for the continent in 1644:

The aspect of the world around him in which he had always sought for beauty was disfigured by violence and death. Hollar therefore had every reason to depart from a country where havoc was rife, and to try and find a refuge elsewhere. His cautious attitude saved him from any direct attacks, but he felt that this immunity might be withdrawn from him at any moment. His past record clearly suggested Royalist sympathies [...] [a]nd the fact he had drawn portraits of [prominent parliamentarians] Milton and Pym availed him little against the accusation of being a Royalist partisan. [...] Although being a foreigner, he was unfamiliar with the problems of English state religion and constitution, and although he had tried to avoid taking sides actively in these matters, he found himself from day to day being more and more forced into a position in which some kind of partisanship would be inevitable.[45]

Similarly, Gillian Tindall infers from Hollar's immigrant status an intended detachment or ambiguity in relation to the politics of the internecine conflict. 'You could surmise that Hollar, as a foreigner,' writes Tindall, 'was most unwilling to take sides in a quarrel among Englishmen that was nothing to do with him'.[46] This was further mitigated by the intensely political, religious and familial sectarian divisions, which engendered a conflict marked by deep social, cultural and even familial dislocation.[47]

Recognition of the ambivalency of Hollar's status is crucial because of its attendant abrogation of straightforward biographical interpretations of the map as a manifestation of personal political expression. Hollar's undetectable allegiance reiterates one of the defining characteristics of his career as a cartographer, namely a pragmatic willingness to produce maps for a particular readership even in times of the explosive conflict. While the map added to the royalist account of a siege suggests an allegiance to the crown during the conflict, his later decorative map to the Fairfax, one of the most prominent parliamentarian generals of the period conveys the sense of a cartographer unconstrained by partisan loyalties in his professional activities.

In the early modern period, maps depicting military conflict were particularly prominent in tracts about Ireland. As we have seen in the previous two chapters in particular, in the cartography of the sixteenth and seventeenth centuries, representations of England's oldest colony were emblematic of wider trends in the publication and dissemination of maps and atlases. 'The seventeenth century proved to be an important formative period in the writing of regional geography,' writes F.V. Emery, 'and Ireland played a distinctive part in its progress from start to finish'.[48] Published in the aftermath of 1641 rising and almost contemporaneous with Hollar's most substantial surveying texts, another anonymous pamphlet entitled 'A true relation of divers great defeats given against the rebells of Ireland, by the Earle of Ormond, lieutenant-generall of all His Majesties forces in that kingdom' provided the reader, alongside a narrative of the rebellion, with 'a map describing the order of a battell lately fought there'.[49]

In the same year, a single-leaf map was published delineating 'Ye Kingdome of Ireland' and incorporating 'particular notes distinguishing the Townes taken, revolted or burnt since the late Rebellion'. Additionally, the surveyor John Woodhouse published a series of books intended as supplementary reading, emphasising the perceived congruence of maps and conflict narratives. The first two sought to delineate geography: *A Guide for Strangers in the Kingdome of Ireland* (1647) and *The Map of Ireland* (1647). The last of the three, *A True Relation of the Bloody Massacres, Tortures, Cruelties, and Abominable Outrages Committed upon the Protestants Proved upon Oath, and Eye-Witnesses* (1647) sought to portray the sectarian tensions and ongoing religious conflicts in the country. Earlier in the century, John Speed produced a map of the country which surveyed the topography of the land and also detailed the beheading of the rebel leaders Shane O'Neill and the Earl of Desmond, a key moment in England's colonial conflict with autochthonous insurgents.[50] In a similar fashion, Hollar's Plymouth map narrates the various events of the siege through detailed insets.

Meeting consumer demand

Alongside demonstrating maps as a canvas for conflict, Hollar's works also exhibit a responsiveness to consumer demand. Representative of the interests of the early modern cartographic enthusiast is an anonymously authored work published in 1650. Seeking to provide the reader with a guide to 'find any place in this map of Europe', the title reveals forcefully the broad and multi-faceted identity of the consumers of cartography in the period:

> A book and map of all Europe with the names of all the towns of note in that known quarter of the world: so that any one of the least capacity, finding the town in the alphabet, shall presently lay his finger upon the town in the map: a work very usefull for all schollars, marchants, mariners, tradesmen, and all that desire to know forreign parts, and especially in these times of warres and commotions that are now in Europe.[51]

Directed towards a wide audience – 'schollars, marchants, mariners, tradesmen, and all that desire to know forreign parts' – the title also articulates the map-reader's broader political and militaristic interests. In Wenceslaus Hollar's maps, we find a corpus of work which evinces a striking sensitivity to the diversity of this wide-ranging readership. The variety of his cartographies, in style, character and purpose, foreground Hollar's range as a map producer. Rigidly systematised, scaled and meticulously ordered delineations of geographic space such as maps of London exemplify the draughtsman's attention to cartographic detail and ability as a surveyor of topographical exactitude.

Elsewhere, Hollar's output evidences his willingness to employ cartographic expertise within more popular mediums. In his map of the Isle of Man dedicated to Fairfax, we discern a graphic artist conversant with the decorative elements of cartographic representation, as well as its diversity of projections from bird's eye view to scenic panoramas to urban views. Additionally, the variety of his mapping scenes powerfully illustrates Hollar's sensitivity to the diverse demands of the early modern map-reader and their multi-disciplinary interests. Overall, Hollar's cartographic oeuvre demonstrates an ability to exploit the medium of cartographic representation to meet a whole host of particular ends – geographic specificity (*Quartermaster Maps*), consciously politicised narratives (Civil War broadside), propagandistic delineations (Devereux portrait) or aestheticised objects for contemplative gratification (Fairfax map of the Isle of Man).

Notes

1. Lionel Cust, 'Notes on the Collections Formed by Thomas Howard, Earl of Arundel and Surrey', *The Burlington Magazine for Connoisseurs*, 19:101 (1911), p. 280. For a more extensive account of Hollar's trip to England, see Francis C. Springell, *Connoisseur and Diplomat: The Earl of Arundel's Embassy to Germany in 1636, with a Transcription of the Diaries of William Crowne* (London: Maggs Bros., 1963). Johannes Urzidil's *Hollar: A Czech Émigré in England* (London: The Czechoslovak, 1942) provides an account of the tempestuous political and social situation in Prague at the time of Hollar's departure (pp. 16–21).
2. Howard quoted in Springell, *Connoisseur and Diplomat*, p. 143.
3. For an indication of the extraordinary range of Hollar's output, the British Museum website has digitised over 2500 drawings contained in its extensive collection of works by the artist, including (but not limited to) portraits, seascapes, caricatures, frontispieces, zoological diagrams and book illustrations. For a discussion of Hollar's important contributions to the work of historian William Dugdale, including engravings for Dugdale's influential *Monasticon Anglicanum* (1655), see John Barnard, 'London Publishing, 1640-1660: Crisis, Continuity, and Innovation', *Book History*, 4:1 (2001), pp. 8–10. The title of Anne Thackeray's study of Hollar's work, *Caterpillars and Cathedrals: The Art of Wenceslaus Hollar* (Toronto: Thomas Fisher Rare Book Library, 2010), is indicative of the artist's range. Matthew C. Hunter describes Hollar as 'arguably the leading graphic artist of the Restoration'. See his essay 'Hooke's Figurations: A Figural Drawing Attributed to Robert Hooke', *Notes and Records of the Royal Society*, 64:3 (2010), p. 256.
4. George Vertue, *A Description of the Works of Wenceslaus Hollar* (London: William Bathoe, 1745), pp. 21–45. Richard Pennington, *A Descriptive Catalogue of the Etched Works of Wenceslaus Hollar 1607–1677* (Cambridge: Cambridge University Press, 1982), pp. 104–426.
5. Pennington, *Descriptive Catalogue*, p. xi.
6. John Aubrey, *Aubrey's Brief Lives*, ed. Oliver Lawson Dick (London: Penguin Classics, 1987), p. 241.
7. Craig Hartley, 'The Young Hollar in Prague: A Group of New Attributions', *Print Quarterly*, 8:3 (1991), p. 264.

8. See, for example, Simon Turner's 'Hollar in Holland: Drawings from the Artist's Visit to the Dutch Republic in 1634', *Master Drawings*, 48 (2010), pp. 73–104. Turner observes:

> What was to prove far more influential to Hollar was not the figurative work of Rembrandt but the Dutch landscape tradition, as practised or promoted by such artists as Claes Jansz Visscher (1587–1652) and Jan van de Velde II (c. 1593–1641). The Northern Netherlands, particularly Amsterdam, was equally renowned for its burgeoning and innovative cartographic publishing adventures, and Hollar's own youthful 'mapping impulse' must have driven him to find out more about the impressive wall maps, atlases, charts and globes produced, among others, by Willem Jansz Blaeu (1571–1638) and Jodocus Hondius 1st (1563–1612) and their successors. (p. 78)

9. Katherine S. Van Eerde, *Wenceslaus Hollar: Delineator of His Time* (Virginia: Folger Books, 1970), p. 8.
10. The British Library website contains a map of Deptford, London, annotated by Evelyn http://www.bl.uk/onlinegallery/onlineex/deptford/m/008add00078629au00000002.html (accessed 21st November 2013). Giles Madelbrote's examination of Evelyn's personal library provides an initial attempt to 'reconstruct [Eveleyn's] methods of study, devotion and work' notably includes extensively annotated copies of Evelyn's own *A Philosophical Discourse of Earth* (1676) and *Sylva* (1679) which dealt with landscape and soils, two recurrent subjects in the broader field of geography in the early modern period, as the recurrence of 'soyl' in the previously cited chorographies of Speed's *Theatre* attest. See Mandelbrote, 'John Evelyn and His Books', *John Evelyn and His Milieu*, eds. Frances Harris and Michael Hunter (London: British Library, 2003), pp. 71–90.
11. P.D.A. Harvey, *Maps in Tudor England* (Chicago: University of Chicago Press, 1993), p. 7.
12. Harvey, *Maps in Tudor England*, pp. 7–8.
13. As Rhonda Lemke Sanford points out, Leo Bagrow's authoritative *History of Cartography*, published in 1963, adopts a similarly broad definition of its subject Sanford, *Maps and Memory in Early Modern England*, p. 2. See Leo Bagrow, *History of Cartography* (Cambridge: Harvard University Press, 1963), p. 22.
14. Sanford, *Maps and Memory in Early Modern England*, p. 3. See also David Woodward, 'Cartography and the Renaissance: Continuity and Change', *The History of Cartography, Volume Three, Part 1: Cartography in the European Renaissance*, ed. David Woodward (Chicago: Chicago University Press, 2007), pp. 3–24.
15. See Miloš V. Kratochvíl, *Hollar's Journey on the Rhine* (Prague: Artia, 1965) for Hollars prolificacy in this genre.
16. Gavin Hollis, '"Give me the map there": *King Lear* and Cartographic Literacy in Early Modern England', *Portolan*, 68 (2007), p. 11.
17. Pennington's survey of Hollar's work provides a detailed summary of the publishing circumstances of this single sheet map. See Pennington, *A Descriptive Catalogue*, p. 186.
18. Donald K. Smith, *The Cartographic Imagination in Early Modern England* (Surrey: Ashgate, 2008), p. 63.
19. David Harvey, 'Cosmopolitanism and the Banality of Geographical Evils', *Public Culture* 12:2 (2000), p. 549.

20. Paige Newmark, '"She is Spherical, Like a Globe": Mapping the Theatre, Mapping the Body', *Shakespeare in Southern Africa*, 16 (2004), p. 17.
21. Jerry Brotton, 'Printing the Map, Making a Difference: Mapping the Cape of Good Hope, 1488–1652', *Geography and Revolution*, eds. David N. Livingstone and Charles W. Withers (Chicago: University of Chicago Press, 2005), pp. 137–159.
22. http://luna.folger.edu/luna/servlet/detail/FOLGERCM1~6~6~1012707~170727:The-kingdome-of-England—principal (accessed 3rd August 2019).
23. John Speed, *The Theatre of the Empire of Great Britaine* (London: William Hall, 1612), sig. ¶ 2r.
24. Monique Pelletier, 'Representations of Territory by Painters, Engineers, and Land Surveyors in France during the Renaissance', *The History of Cartography, Volume Three, Part 1: Cartography in the European Renaissance*, ed. David Woodward (Chicago: Chicago University Press, 2007), p. 1532.
25. Ralph Hyde, *Gilded Scenes and Shining Prospects: Panoramic Views of British Towns, 1575–1900* (New Haven, Connecticut: Yale Center for British Art, 1985), p. 11. See also Sanford, *Maps and Memory in Early Modern England*, p. 108.
26. Pennington, *Descriptive Catalogue*, p. 115.
27. Simon Turner, 'Hollar's Prospects and Maps of London', *Printed Images in Early Modern Britain: Essays in Interpretation*, ed. Michael Hunter (Surrey: Ashgate, 2010), p. 147.
28. Wenceslaus Hollar quoted in Gillian Tindall, *The Man Who Drew London: Wenceslaus Hollar in Reality and Imagination* (London: Vintage, 2002), p. 143.
29. Albert C. Labriola, 'Painting and Poetry of the Cult of Elizabeth I: The Ditchley Portrait and Donne's "Elegie: Going to Bed"', *Studies in Philology*, 93:1 (1996), pp. 42–49.
30. For a discussion of the proliferation of broadsides such as Hollar's, see Renata Shaw, 'Broadsides of the Thirty Years' War', *The Quarterly Journal of the Library of Congress*, 32:1 (1975), pp. 2–24.
31. Kratochvil, *Hollar's Journey on the Rhine*, p. 11.
32. J.B. Harley, *The New Nature of Maps: Essays in the History of Cartography*, ed. Paul Laxton with an introduction by J.H. Andrews (Baltimore and London: Johns Hopkins University Press, 2001), p. 57.
33. Anonymous, *A True Narration of the Most Observable Passages, in and at the Late Seige of Plymouth, from the Fifteenth Day of September 1643, untill the Twenty Fift[h] of December Following* (London: L. N., 1544), sigs. C2r-2v. Pennington attributes authorship of the work to its printer, John White. See Pennington, *Descriptive Catalogue*, p. 88.
34. For a more extensive day-by-day account of Maurice's siege of Plymouth in late 1643, see Wilfred Emberton, *The English Civil War Day By Day* (1997), pp. 80–92, and Mary Coate, *Cornwall in the Great Civil War and Interregnum, 1642–1660: A Social and Political Study* (Truro: Barton, 1963), pp. 164–170. J. R. Powell's *The Navy in the English Civil War* (Hamden: Archon Books, 1962) provides a context for Maurice's attempts to capture the city within the overall events of the war (pp. 46–55).
35. Anonymous, *Good nevves from Plymouth Being a True Relation of the Death of Sir Ralph Hopton, and Many of His Commanders* (London: 1643), sig. A1r.
36. Anonymous, *A Letter from Plymouth Concerning the Late occurrances and affaires of That Place* (London: Printed for Leonard Smith, 1644), sig. A2r.
37. Anonymous, *Joyfull nevves from Plimouth* (London: Printed by L. N. for Francis Eglesfeild, 1644), sig A1r.

38. Edward W. Said, *Culture and Imperialism* (London: Vintage, 1994), p. 7.
39. Michael Wintle, 'Renaissance Maps and the Construction of the Idea of Europe', *Journal of Historical Geography*, 25:2 (1999), p. 137.
40. Pennington, *Descriptive Catalogue*, p. 88.
41. Antony Griffiths and Gabriela Kesnerová, *Wenceslaus Hollar: Prints and Drawings* (London, British Museum Publications, 1983), p. 31.
42. Robert J. D. Harding, 'Hollar, Wenceslaus (1607–1677)', *ODNB* (Oxford: Oxford University Press, 2004), online edn. January 2008, http://www.oxforddnb.com/view/article/13549 (accessed 20th December 2013).
43. Hind, *Wenceslaus Hollar*, p. 4.
44. Urzidil, *Hollar: a Czech émigré in England*, p. 39.
45. Urzidil, *Hollar*, pp. 41–42.
46. Tindall, *The Man Who Drew London*, p. 48.
47. See Wilfred Emberton, *The English Civil War Day by Day*, p. xvi.
48. F.V. Emery, 'Irish Geography in the Seventeenth Century', *Irish Geography*, 3:5 (1958) p. 263.
49. Anonymous, *A True Relation of Divers Great Defeats Given against the Rebells of Ireland, by the Earle of Ormond, Lieutenant-Generall of All His Majesties Forces in That Kingdom* (London: Robert Barker, 1642), sig. A1r.
50. Patricia Palmer, *The Severed Head and the Grafted Tongue: Literature, Translation and Violence in Early Modern Ireland* (Oxford: Oxford University Press, 2014), p. 31. For a more extensive examination of this type of hybridised narrative/map literature see Emery, 'Irish Geography in the Seventeenth Century', pp. 263–276; J.H. Andrews, 'Science and Cartography in the Ireland of William and Samuel Molyneux', *Proceedings of the Royal Irish Academy*, 80C (1980), pp. 231–250; and Stan Mendyk, 'Gerard Boate and *Irelands Naturall History*', *The Journal of the Royal Society of Antiquaries of Ireland*, 115 (1985), pp. 5–12.
51. Anonymous, *A Book and Map of All Europe with the Names of All the Towns of Note in That Known Quarter of the World* (London: James Moxon, 1650), A2r.

Conclusion

Mapping the stars. And the future

On the 19th December 2013, the British Broadcasting Corporation (BBC) reported on its news website on the launch of the Gaia satellite by the European Space Agency (ESA) from the coastal town of Sinnawary in French Guiana. According to the BBC story, the satellite is tasked with a precise mission. 'Gaia', reports the BBC's science correspondent Jonathan Amos, 'is going to map the precise positions and distances to more than a billion stars. [...] It will be engaged in what is termed astrometry – the science of mapping the locations and movements of celestial objects'.[1] Similarly, in 2018 the same website reported on the launch of the ESA's Aeolus satellite, describing how it is designed to map the earth's wind patterns from space.[2]

Both the Gaia Project and Aeolus continue a long tradition of cosmically orientated exploration stretching back millennia. For example, the layout of many prehistoric buildings, tombs and monuments bears proof of an observant astronomical gaze, with architectural features evidencing rigorous attention to the movements of the stars from the very beginning of human civilisation.[3] According to John North, 'It is no exaggeration to say that the astronomy has existed as an exact science for more than five millennia'.[4] Writing in the preface to his survey *Cosmos: An Illustrated History of Astronomy and Cosmology*, North continues:

> Writing of [astronomy's] history presents us with innumerable problems. We begin with a period known to us largely by inference; we continue into times from which much of the evidence is known to have been lost; and we end with the last decades of a century that has provided astronomers with unprecedented attention and economic resources. From a typical century in the Hellenistic period, a golden era of astronomy, we might be left with a mere handful of texts. By contrast, there are now more than 20,000 astronomical articles published every year, and over a five-year period the number of astronomers under whose names they appear is of the order of 40,000.[5]

North's observations on the history of astronomy underline the historical duration of the subject. From a period of obscurity to one of almost

DOI: 10.4324/9781003200376-12

ubiquity, the narrative of astronomy from prehistoric to the contemporary appears ineluctably teleological. Yet, while the relatively recent increase in astronomical literature suggests a particular burgeoning of the science, between the convenient statistical bookends mentioned by North – the Ancient Greek world and the modern world of the twentieth and twenty-first centuries – several epochs stand out as 'golden ages' of cosmological exploration, not the least of which is the sixteenth and seventeenth century.

Plotting the stars

Within this period, astronomical endeavours increased exponentially as early modern stargazers drew heavily on the work of ancient Greco-Roman and more recent Arabic precursors. The translation and proliferation of works such as Claudius Ptolemy's *Almagest* (c.127–141 AD) in particular proved immensely influential, serving as a catalyst for a renewed interest in the cosmos.[6] In Persia, following on from the advances of Abd al-Rahman al-Sufi whose seminal *Book of Fixed Stars* (c.964) set out the trajectories, magnitude and colours of many of the most visible stars in the night sky, later figures such as Ulugh Beg, Ibn al-Haytham and Taqī al-Din Muhammad ibn Ma'ruf began compiling their own stellar catalogues, setting out a map of the stars and their movements.

As Thomas F. Glick, Steven Livesey and Faith Wallis explain, Ibn al-Haytham's work on vision and light, principal cornerstones of astronomical science, 'constitut[e] the most remarkable accomplishment in optics from the times of Ptolemy to those of Johannes Kepler'.[7] With its citation of the work of the German mathematician Johannes Kepler, Glick, Livesey and Wallis point towards the substantial involvement of Renaissance Europe in the development of astronomy in the post-medieval age. Awash with cosmographical texts translated from classical sources as well as those from Arabic and Asian astronomers and contemporary writers, European intellectualism maintained and sustained astronomical investigations of its own. Nicolaus Copernicus's seminal *De revolutionibus orbium coelestium* ('On the Revolutions of the Heavenly Spheres'), published in 1543, sought to address and ultimately refute the established geocentric models of the universe that prevailed since the time of Ptolemy. Copernican astronomy introduced a heliocentric model which would be debated, often violently, over the next three centuries: famously, the Italian polymath Giordano Bruno would be burnt at stake by the Catholic Church for, among other charges, propagating heliocentrism.[8] In Denmark, the polymath Tycho Brahe, emboldened by advances in instrumental technology and especially the refinement of telescopes, had begun to set out a more expansive system of co-ordinates of constellations. Later the work of Brahe's assistant Kepler would synthesise astronomy and physics to theorise the first laws of planetary motion and as a consequence, expand the limits of a mathematical understanding of the universe.

Attempts to map the sky ran parallel to and were frequently intertwined with the mapping of the earth. Study of the heavens, as the cosmographical investigations of Christopher Marlowe's Faustus testify, carried a strain of self-reflexivity, turning back on the cosmographical surveyor and their own perception of themselves, their maker and the relative positions of both on the metaphysical schema. Malgorzata Grzegorzewska notes:

> Interpretation gave way to appropriation: universe explained and described by man became his property rather than his image and likeness [.] [...] Trapped between two distinct versions of the world-as-book concept, between image and Euclidean line, maps became a handy tool for encompassing and domesticating the universe. At the same time, mapmaking and readership were inevitably caught in the network of imaginary travels to imaginary lands.[9]

Mapping and astronomical investigation, conflated in the publicity spiel of the Gaia Project, synergised in the febrile scholarly environment of early modern England. Many writers and intellectuals cultivated an interest in the movements of the stars and planets, reflecting at once the central place afforded the subject of astronomy in the education curriculum as part of the quadrivium alongside arithmetic, geometry and music and also the prevailing development of map-mindedness.[10]

Combining theory and practice

In keeping with an interdisciplinary scope which sought both theoretical and practical applications, writers and intellectuals explored both the speculative and experimental aspects of astronomy. Many embodied the close relationship between mapping and exploration in both the terrestrial and heavenly spheres. John Dee discussed throughout this book for his observations on the delight, purpose and edification of map reading, for example, may be regarded as one of the foremost English astronomers of the sixteenth century. In addition to engaging in astrological prognostication and composing star charts plotting the cosmos, Dee, alongside Thomas Digges (like Bruno, a proponent of Copernican ideas) 'cherished an ambitious programme of reforming the whole of astronomy' based upon forms of observation which challenged orthodoxy.[11] Later, Robert Burton incorporated astronomical investigation into his magnum opus, *The Anatomy of Melancholy*: in a section entitled 'Digression of Air', the writer and anatomist of psychological ailments presents an extensive deliberation on the nature of the universe and the possibility of multiple worlds beyond the terrestrial.[12] Francis Bacon's vast erudition and interests also extended to cosmology and cosmography. In his planned, though unfinished, work *The Great Instauration* (1620), the putative father of modern science attempted to postulate a cosmological system of his own based upon a finite universe with a geocentric structure.[13]

As with the cartographies practiced by figures such as Ortelius, Saxton and Speed, these endeavours percolated into social and cultural discourses beyond the scientific. Famously, William Shakespeare's *King Lear* includes its own mapping moment, with the titular protagonist summoning cartographic representation in a fashion similar to Tamburlaine.[14] Yet the play also contains an extensive soliloquy on the nature of early modern astronomical thought and its closely related pursuit of astrological divination. About halfway through act one, scene two, the illegitimate son of the Duke of Gloucester and villain of the piece, Edmund cogitates on the vanity of prediction by cosmological activity:

> This is the excellent foppery of the world that when we are sick in fortune - often the surfeit of our own behaviour - we make guilty of our disasters the sun, the moon, and the stars, as if we were villains by necessity, fools by heavenly compulsion, knaves, thieves, and treachers by spherical predominance, drunkards, liars, and adulterers by an enforced obedience of planetary influence, and all that we are evil in by a divine thrusting-on. An admirable evasion of whoremaster man, to lay his goatish disposition to the charge of a star! My father compounded with my mother under the dragon's tail and my nativity was under Ursa Major, so that it follows I am rough and lecherous. Fut, I should have been that I am, had the maidenliest star in the firmament twinkled on my bastardizing.[15]

In hearing Edmund's disdain for 'spherical dominance' and 'whoremaster man['s]' servility to it, we witness a refutation of the validity of astrology. Concomitantly, Edmund tacitly voices an acknowledgement of horoscopy's prevalence in seventeenth-century England and the ways in which it sought, like Donne's surveying net discussed at the start of this study, to create a map-like order of the seemingly disordered universe.

Furthermore, Edmund powerfully portrays a responsivity to dominant intellectual trends in much the same manner as the literary reactions to maps. The parallels with Puttock's dismissal of cartography's 'prittie painted things' and enunciated rejection of the world-on-world-paradigm entertained by the map enthusiast are palpable. Yet running through both confutations is an implicit and underlying acknowledgement of the popularity and pervasiveness of such ideas in the wider contemporary intellectual community. In spite of the contemptuous rhetoric of Puttock and Edmund, both characters attest to the fact that maps in the mind of the early modern map reader, whether of the earth or the heavens, retained an ability in the early modern consciousness to supersede a purely representational function.

The poet and the astronomer

As with drama, poetry often took recourse to the cosmological mappings of the age. Having visited Galileo Galilei during his tour to the continent in 1638, John Milton made the Italian polymath the sole contemporary figure in his epic poem *Paradise Lost* (1667).[16] Milton lyrically confers an exalted intellectual status upon Galileo, most especially in alerting us to his contemporary's cartographically inclined endeavours. Describing the great hero of the poem Satan as he moves across the 'burning marl' of hell after his initial defeat, the poet trains his gaze on the fallen archangel's armaments. Citing the 'Tuscan artist' Galileo, Milton articulates one of his most renowned epic similes:

> He scarce had ceased when the superior Fiend
> Was moving toward the shore, his ponderous shield,
> Ethereal temper, massy, large, and round,
> Behind him cast; the broad circumference
> Hung on his shoulders like the moon, whose orb
> Through optic glass the Tuscan artist views
> At evening, from the top of Fesole
> Or in Valdarno to descry new lands,
> Rivers, or mountains in her spotty globe.
>   (Book 1, 283–291)[17]

Milton's Galileo enacts a strikingly analogous undertaking to those behind the Gaia Project's attempt to 'make [...] [a] multidimensional map' of the universe. The Italian astronomer is figured as a topographical surveyor, 'descry[ing] new lands' in the firmament. The sense of novelty attending the actions of the Miltonic Galileo conveyed through the indefiniteness of the conjunction 'or' in the middle of an outline of the moon's topography and the pointed use of the word 'new' in the phrase 'new lands', alert us to the undiscovered and unknown at the heart of the enterprise. Nearly five hundred years after the 'Tuscan artist' sat in twilight charting the unfamiliar lunar landscape, the Gaia Project will maintain astronomy's surveying of new and undiscovered lands.

The astronomical and astrological presence in *King Lear* and *Paradise Lost* corroborate not only the prevalence of surveying in the early modern period. They also reveal how artistic mediums can interpret scientific pursuits, representing, staging, reflecting and even challenging their methodologies. They serve as a backdrop for considering how mapping technologies function. The rhetoric of the Gaia Project reiterates the ostensible purpose of cartography as a medium of calibrating, systematising and disseminating information. In the stated objective of the Gaia Project to 'make the largest, most precise three-dimensional map of our Galaxy by surveying more than a thousand million stars' the language of the cartographer is invoked – the astronomers will produce a 'map' having 'surveyed' the sky. Examining both the attempt by early modern mappers to map the topography of the terrestrial; and their striving of the Gaia Project to capture cartographically the extraterrestrial, we find many connexions between their day and ours. According to Denis Cosgrove, astronomy was a participant in a discourse of control and 'mastery' that was characteristic of cartography as a whole:

> In the Renaissance images played an important role in remapping medieval natural philosophy. Renaissance cosmography might be regarded as a 'mode,' or a historically specific set of social and technical relations that determine representational practices. The social and technical relations of Renaissance cosmography converged around a growing apprehension of terrestrial, celestial, and representational space as absolute and capable of intellectual mastery.[18]

Evoking John Donne's observation on Renaissance mapping endeavours that 'of meridians and parallels | man hath weaved out a net', Cosgrove draws our attention to the intersecting motivations of the astronomer, cartographer and cosmographer. There is a continual and recognisable inclination towards cartographic conceptualisation when viewing the novel and to-be discovered: maps, geographical and cosmographical, 'factual' and prognosticatory, 'realist' and allegorical from the age of Marlowe, Shakespeare and Middleton to our own retain their pliability in the pursuit of ingesting, computing and displaying information. There persists an attempt to not only familiarise the unknown but also exert control over it for specific ends.

Walking the Milky Way

Meaningfully for our investigation into the points of interaction between cartographies intellectual and imaginative, the Gaia Project reiterates another concept buttressing cartographic enthusiasm in early modern England, namely the notion of travelling by map. Thus, the attempted cartographic construction of an image of the cosmos by the Gaia Project is lauded fulsomely by Gerry Gilmore, Professor of Experimental Philosophy at the University of Cambridge and Fellow of the Royal Society. Gaia's map, Gilmore enthuses, 'will allow us, for the first time ever, to walk through the Milky Way - to say where everything is, to say what everything is'.[19] Such words strike a plangent note when read in the context of the ideas, concepts and impulses of Renaissance cartography. There is much in Gilmore's effusion that is reminiscent of the early modern mapping aficionados: as this study has attempted to demonstrate, the idea of travelling by map was popular among a whole host of writers, pervading the imagination and manifesting articulations in prose, poetry and drama as well as didactic, political and socio-economic debates. In language redolent of his fellow Oxbridge don Robert Burton who 'never travelled but in map or card', Gilmore reminds us again of the seeming power of maps to transcend the constraints of the immediately physical. As such, maps, even in an age characterised by cartographic exactitude initiated by satellite surveys and Google Maps, retain their appeal as more than unmediated representations of landscapes, seascapes and skyscapes. They can also function as agents, or to borrow Jonathan Sawday's eloquent description of early modern technologies, as 'engines' of the imagination, both firing and being produced by our mental faculties. 'Just like our own machines', Sawday writes:

> Renaissance machines were useful devices with which people worked and laboured, Acting upon the world, their avowed purpose was to make human existence more tolerable. But fabricated as they were out of a synthesis of poetry, architecture, philosophy, antiquarianism, and theology, as well as craft, skill, and design, Renaissance machines were also freighted with myth, legend, and symbolism.[20]

The freighting of machines with social, political and cultural discourses extended, as we have seen, to the technologies of cartography. Contrary to Ralph Hertel's claim that cartography 'symbolizes a translation of place as it is subjectively experienced into a more neutral form of space',[21] the practice of surveying and setting down of topography onto a two-dimensional plane represents an enterprise which draws upon the integument of objectivity and neutrality to propagate a particular subjective perspective. Early modern maps and, more broadly, pictorial representations of geographies were ostensibly 'real', but they were also politicised, allegorised and poeticised.

One of the key effects of this meeting of real and created was the idea of 'vicarious travel'. The symbioses outlined by Sawday – between techne and mythos, machine and mind – lie at the heart of the Renaissance engagement with the map, with the notion of imaginative travelling serving as one of a number of intentions behind cartographic production and consumption. Like Gilmore's envisaged 'walk through the Milky Way', Marlowe's Tamburlaine travels through his demesne via a mapped representation of the world drawn from Ortelius's surveys; John Norden's landowner looks upon the estate survey and 'may see what he hath, where, and how it lieth, and in whose use and occupation every particular is, upon the suddaine view;[22] and Robert Herrick's brother sails to foreign and exotic locales while ensconced in his country house. As Gilmore's intention to travel through the galaxy on the back of the mapping exploits of the Gaia satellite proves, the interaction between imagination and science in the attempt to travel by map persists and thrives in our own time.

Surveying the political

Another twenty-first-century mapping endeavour serves to underline the similarities between the cartographies of the current age and those of the early modern period: the many applications and appropriations of cartographically related disciplines and ideas. In May 2014, the Glasgow-based think tank The Jimmy Reid Foundation issued a volume from its latest press entitled *An Atlas of Productivity*.[23] Produced in collaboration with the Glasgow-based research and design collective Lateral North, this elaborate and handsome-produced publication pointedly styled itself as 'the first dedicated atlas of Scotland since the 19th century and perhaps the first 'atlas of productivity' anywhere in the world'. It forms parts of the graphic design collective's '"boutique" collection of Scotland-centric cartographic works'.[24] Reminding us of the multifaceted nature of the cartographic gaze, and the ease of its usage as a medium of a wide range of information, the atlas attempts to map:

> not just Scotland's landscape and towns and cities but seeks to map as many of the aspects of national productivity as possible. Over 35 maps we see everything from Scotland's land ownership and its wind speeds to its transport links and its relationship to the emerging Arctic trading routes.[25]

In such rhetoric, there is a conspicuous awareness of the ways in maps can be used to express, project and compute a panoramic range of diverse data. Patently, the 'productivity' designation of the *Atlas*'s title serves to characterise the text as more than a survey of the Scottish landscape. However, in attempting to map 'land ownership', 'wind speeds', 'transport links' and even a wider network of international maritime trade, the *Atlas*'s rubric evidences a sensitivity to how mapping methodologies may be employed in the dissemination of non-topographical information. In a striking parallel to the many mapping moments examined, from Edmund Spenser's colonial 'plot' to William Cuningham's cosmographical 'glasse' to Henry Peacham's pedagogical aids, the authors of *An Atlas of Productivity* demonstrate how cartography can be used as a medium of demographic, economic and meteorological as well as topographical material.

Emerging some four months before a key moment in the history of the United Kingdom – the referendum for Scottish independence of 2014 – its timeliness recalls another atlas incipient at a crucial formational moment for notions of British archipelagic identities: John Speed's *Theatre of the Empire of Great Britaine* (1612). The aesthetic parallels between the *Atlas of Productivity* and Speed's *Theatre* are especially striking. Handsomely produced, the physical scale of *The Atlas of Productivity* recalls Speed's weighty, plus-500 folio pages publication. Moreover, like the earlier text, which provides a chorographic-cartographic hybrid, Lateral North provides maps opposite data and information regarding Scotland's economic climate, as on page 21, which sets out figures for Scotland's oil and gas reserves opposite a map of among other details, the country's exporting and importing pipelines.

Setting out both the purpose of the publication and the manifesto of its authors, we gain an insight into the internal dynamics of the project and the factors driving its production. 'Lateral North', it states:

engages in cross-sectoral, multidisciplinary design and research. [This] collaboration encourages discover, investigation and testing of previously overlooked opportunities and seeks to engage in long-term strategic and holistic visioning.[26]

The interdisciplinary scope is discernible in the terminology of the introduction. It continues:

Our integrated design approach involves extended engagement alongside creative professionals and attempts to deliver a competent design resolution for projects ranging from historic site conservation and community trust developments to exhibition and graphic design.[27]

Phrases such as 'cross-sectoral', multidisciplinary', 'collaboration', 'integrated approach', extended engagement, and especially 'holistic visioning' at once flag up the collaborative and inclusive character of the cartographically inflected project; and simultaneously echo works such as Speed's *Theatre*. As witnessed, early modern mappers interacted with, employed and even corralled a whole host of epistemologies and intellectual endeavours, from antiquarianism to etymology to climatology, in the production of a particular image of Britain and the respective statuses of England, Wales, Ireland and Scotland within the nascent imperial entity. Likewise, the authors of *An Atlas of Productivity* utilise expertise from fields as diverse as architecture, demography and economics to present a cartographic image of its subject. Once more, mapping's cross-pollination with other, superficially discrete disciplines – a major hallmark of both early modern cartography and the early modern perception of cartography – is in evidence.

In addition, the Reid Foundation's document illustrates the ready applicability of cartography to socio-political discourse. It exemplifies the ways in which maps can be appropriated as a medium not only of topographical representation, but also a medium of dissemination of broader ideas regarding the identity of self and nation. Speed's *Theatre* was suffused with unificatory rhetoric, employing the vocabulary of the anatomist to project an image of a naturally ordered, precisely constituted and inherently hierarchised nation and imperial state. Similarly, the more recent cartographic publication is imbued with a subtle though unmistakeably political motivation: attempting to project an image of an independent Scotland's economy in the twenty-first century, it draws the country (quite literally) within a series of carefully calibrated landscapes. 'Lateral North', states the introduction, 'looks to investigate Scotland's new place and identity within an economically emerging northern region, exploring the relationship between people, culture, places, industries and economies'.[28]

Confecting the land

As with John Speed's *Theatre,* which maps all four nations of the British isles under the one national and imperialist umbrella of 'Great Britaine', what the mapper elects to map (and as a corollary, what not to map) is significant. The *Atlas* proffers to the reader three different cartographic contextualisations of Scotland, depicting the country firstly on its own from the Borders in the south to the Orkney Islands in the north (pp. 11–27); secondly within a northern European context, projecting an image of Scotland as part of Britain but also a wider theatre incorporating Scandinavia, Iceland and Germany (pp. 30–39); and finally within an Arctic 'circle', centring on the North Pole and showing Scotland in relation to the northern coastlines of the American and Eurasian continental landmasses (pp. 40–50). Moving from microcosm to macrocosm along a specific line, the cartographic gaze of this particular atlas decentres the economic and political focal points from the normative 'British' hub of socio-cultural identity.

As these two atlases, standing sentinel at moments of consolidation and possible disintegration of the British nation state prove, cartography is continually co-opted into particular, subjective discourses. Employing what appears at first to be an impartial descriptive mode, mappers frequently appropriate this ostensible neutrality of cartography to project their own highly subjective perceptions of the way the word is and, indeed, occasionally advance how it should be. While Speed presents the limbs of early seventeenth-century Britain as a constituent of an organic whole, the atlas produced by Lateral North advances an image of one of the appendages in an entirely different setting, relocating it within another set of putatively interrelated organs.

Ultimately what the astrometry of the Gaia Project and Aeolus as well as the economically inflected *Atlas of Productivity* attest to is the persistent responsiveness of cartography to the presentation, propagation and consumption of information, ideas and ideologies. In a multitude of ways, maps reveal as much about the mapper and the map reader as they do about the mapped. And this, as I hope to have shown, is exemplified by the many engagements with cartography in early modern English culture.

Notes

1. http://www.bbc.co.uk/news/science-environment-25426424 (accessed 16th January 2021).
2. https://www.bbc.co.uk/news/av/science-environment-44427377 (accessed 8th February 2021).
3. See Clive Ruggles and Michael Hoskin, 'Astronomy Before History', *The Cambridge Concise History of Astronomy*, ed. Michael Hoskin (Cambridge: Cambridge University Press, 1999), pp. 1–17.
4. John North, *Cosmos: An Illustrated History of Astronomy and Cosmology* (Cambridge: Cambridge University Press), p. xxii.
5. North, *Cosmos*, p. xxii.

6. For a brief précis of the spread of Ptolemy's ideas in the west from the medieval period onwards, see Michael Hoskin and Owen Gingerich, 'Medieval Latin Astronomy', *The Cambridge Concise History of Astronomy*, ed. Michael Hoskin (Cambridge: Cambridge University Press, 1999), pp. 68–93. *Ptolemy's 'Geography' in the Renaissance*, ed. Zur Shalev and Charles Burnett (London: Warburg Institute Colloquia, 2011) presents an in-depth series of essays considering the impact of Ptolemaic ideas in Europe in particular.
7. *Medieval Science Technology and Medicine: An Encyclopedia*, ed. Thomas F. Glick, Steven Livesey, Faith Wallis (London: Routledge, 2014), p. 239.
8. For a full transcript of the charges laid against Bruno, including the accusation his writings 'contained many heresies and errors', see Ingrid D. Rowland, *Giordano Bruno: Philosopher / Heretic* (Chicago: University of Chicago Press, 2009), pp. 287–290.
9. Malgorzata Grzegorzewska, '*Theatrum Orrbis Terrarum* on the Court Stage', *Shakespeare and His Contemporaries: Eastern and Central European Studies*, ed. Jerzy Limon, Jay L. Halio (Delaware University of Delaware Press, 1993), p. 220.
10. For an overview of the development of English education from the medieval period to the early sixteenth century, see Nicholas Orme, *Medieval Schools: From Roman Britain to Renaissance England* (New Haven: Yale University Press, 2006), pp. 51–86. According to Ann Moyer:

The study of the heavens, then, still formed part of the general studies of philosophy and liberal arts that led to the bachelor's degree. It also held particular and more specialized interest for the medical faculty, because of the fields' perceived abilities to account for celestial influences on health. An expert physician's training and practice included astrology to some degree; the careers of practicing physicians kept astrological practice active in cities and courts. Important courts, in fact, increasingly retained prominent physician/astrologers, a custom established significantly earlier. Astrological practice could also be found at lower social levels, of course, just as other aspects of medical education and practice were not the exclusive province of university-trained experts.

See Moyer, 'The Astronomers' Game: Astrology and University Culture in the Fifteenth and Sixteenth Centuries', *Early Science and Medicine*, 4:3 (1999), p. 229.
11. R. Goulding, *John Dee: Interdisciplinary Studies in English Renaissance Thought* (Dordrecht: Springer, 2006), p. 42.
12. For a more extensive examination of Burton's astronomical activities, see Richard G. Barlow, 'Infinite Worlds: Robert Burton's Cosmic Voyage', *Journal of the History of Ideas*, 34:2 (1973), pp. 291–302.
13. For a discussion of Bacon's astronomical thought and its reaction to the post-Copernican cosmological view, see Graham Rees, 'Francis Bacon's Semi-Paracelsian Cosmology', *Ambix*, 22 (1975), pp. 165–173 and 'Bacon's speculative philosophy', *Cambridge Companion to Francis Bacon*, ed. Markku Peltonen (Cambridge: Cambridge University Press, 1996), pp. 124–131.
14. *King Lear*, 1.1. 37.
15. *King Lear*, 1.2. 116–130.
16. Barbara K. Lewalski, *The Life of John Milton* (Massachusetts: Blackwell Publishing, 2003), pp. 93–94.
17. *Paradise Lost*, ed. David Scott Kastan (Indiana: Hackett Publishing Ltd., 2005).

18. Denis Cosgrove, 'Images of Renaissance Cosmography, 1450–1650', *The History of Cartography, Volume Three, Part 1: Cartography in the European Renaissance*, ed. David Woodward (Chicago: Chicago University Press, 2007), p. 55.
19. http://www.bbc.co.uk/news/science-environment-25426424 (accessed 16th January 2019).
20. Jonathan Sawday, *Engines of the Imagination: Renaissance Culture and the Rise of the Machine* (London: Routledge, 2007), p. 1.
21. Ralph Hertel, *Staging England in the Elizabethan History Play: Performing National Identity* (Surrey: Ashgate, 2014), p. 42.
22. John Norden, *The Surveyor's Dialogue (1618): A Critical Edition*, ed. Mark Netzloff (Surrey: Ashgate, 2010), p. 25.
23. Graham Hogg et al., *An Atlas of Productivity*, Lateral North (Glasgow: Clydeside Press, 2014), p. 3.
24. https://lateralnorth.com/portfolio/atlas-of-opportunity/ (accessed 16th May 2021).
25. http://reidfoundation.org/2014/05/an-atlas-of-productivity/ (accessed 16th June 2016).
26. *An Atlas of Productivity*, p. 3.
27. *An Atlas of Productivity*, p. 3.
28. *An Atlas of Productivity*, p. 3.

Bibliography

Adorno, Theodore and Horkheimer, Max. 'The Concept of Enlightenment', *Dialectic Enlightenment: Philosophical Fragments*, ed. Gunzelin Schmid Noerr, trans. Edmund Jephcott (California: Stanford University Press, 2002), pp. 1–34.

Adrian, John M. *Local Negotiations of English Nationhood, 1570–1680* (Basingstoke: Palgrave Macmillan, 2011).

Agnew, John A. and Livingstone, David N. eds. *Human Geography: An Essential Anthology* (Oxford: Blackwell Publishing, 1996).

Albano, Caterina. 'Visible Bodies: Cartography and Anatomy', *Literature, Mapping and the Politics of Space in Early Modern England*, eds. Andrew Gordon and Bernhard Klein (Cambridge: Cambridge University Press, 2001), pp. 89–106.

Alexander, Gavin. 'Sir Philip Sidney's *Arcadia*', *The Oxford Handbook of English Prose 1500–1640*, ed. Andrew Hadfield (Oxford: Oxford University Press, 2013), pp. 219–234.

Alpers, Svetlana. *'The Mapping Impulse in Dutch Art', Art and Cartography: Six Historical Essays*, ed. David Woodward (Chicago: University of Chicago Press, 1987), pp. 51–96.

Altizer, Alma B. *Self and Symbolism in the Poetry of Michelangelo, John Donne and Agrippa D'Aubigne* (The Hague: Springer, 1973).

Andrews, J.H. 'Science and Cartography in the Ireland of William and Samuel Molyneux', *Proceedings of the Royal Irish Academy. Section C: Archaeology, Celtic Studies, History, Linguistics, Literature*, 80C (1980), pp. 231–250.

— *Plantation Acres: An Historical Study of the Irish Land Surveyor and His Maps* (Belfast: Ulster Historical Foundation, 1985).

— *Shapes of Ireland: Maps and Their Makers 1564–1839* (Dublin: Geography Publications, 1997).

— 'John Norden's Maps of Ireland', *Proceedings of the Royal Irish Academy. Section C: Archaeology, Celtic Studies*, 100C:5 (2000), pp. 159–206.

— 'Statements and Silences in John Speed's Map of Ulster', *The Journal of the Royal Society of Antiquaries*, 138 (2008), pp. 71–79.

Anglicus, Bartholomaeus. *On the Properties of Things in John Trevisa's Translation of Bartholomeus Anglicus, De proprietatibus rerum. A Critical Text*, vol. 3 (Oxford: Clarendon Press, 1988).

Anonymous. *A True Relation of Divers Great Defeats Given Against the Rebells of Ireland, by the Earle of Ormond, Lieutenant-Generall of All His Majesties Forces in That Kingdom* (London: Robert Barker, 1642).

—. *Good Nevves from Plymouth Being a True Relation of the Death of Sir Ralph Hopton, and Many of His Commanders* (London: Printed by L. N. for Francis Eglesfeild, 1643).

— *A True Narration of the Most Observable Passages, in and at the Late Seige of Plymouth, from the Fifteenth Day of September 1643, Untill the Twenty Fift[h] of December Following* (London: L. N., 1544).

— *A Book and Map of All Europe with the Names of All the Towns of Note in That Known Quarter of the World* (London: James Moxon, 1650).

Appelbaum, Robert. 'Utopia and Utopianism', *The Oxford Handbook of English Prose 1500–1640*, ed. Andrew Hadfield (Oxford: Oxford University Press, 2013), pp. 253–267.

Aristotle. *Aristotles Politiques, or Discourses of Gouernment, Transl. Louis Le Roys* (London: Adam Islip, 1598).

Avery, Bruce. 'Mapping the Irish Other: Spenser's *A View of the Present State of Ireland'*, *ELH*, 57:2 (1990), pp. 263–279.

Bacon, Francis. *The Historie of the Reigne of King Henry the Seuenth* (London: Printed by W.G. for R. Scot, T. Basset, J. Wright, R. Chiswell, and J. Edwyn, 1676).

— *The History of the Reigns of Henry the Seventh, Henry the Eighth, Edward the Sixth, and Queen Mary the First* (London: 1676).

— *The Essays*, ed. with an Introduction by John Pitcher (London: Penguin Books, 1985).

— *The Major Works*, ed. Brian Vickers (Oxford: Oxford University Press, 2002).

Bagrow, Leo. *History of Cartography* (Cambridge: Harvard University Press, 1963).

Baker, J.N.L. 'Academic Geography in the Seventeenth and Eighteenth Centuries', *Scottish Geographical Magazine*, 51:3 (1955), pp. 129–144.

Baker, David J. 'Off the Map: Charting Uncertainty in Renaissance Ireland', *Representing Ireland: Literature and the Origins of Conflict 1534–1660*, eds. Brendan Bradshaw, Andrew Hadfield and Willy Maley (Cambridge: Cambridge University Press, 1993), pp. 76–92.

Baldwin, R.C.D. *'Nicholls, Thomas (1532–1601)'*, *ODNB* (Oxford: Oxford University Press, 2004), online edn. January 2008, http://www.oxforddnb.com/view/article/20124 (accessed 21st February 2019).

Barber, C.L. 'The Form of Faustus' Fortunes Good or Bad', *The Tulane Drama Review*, 8:4 (1964), pp. 92–119.

Barber, Peter. *'Mapmaking in England, ca. 1470–1650' The History of Cartography*, Volume 2, ed. David Woodward (Chicago: University of Chicago Press, 2007), pp. 1589–1659.

Barnard, John. 'London Publishing, 1640–1660: Crisis, Continuity, and Innovation', *Book History*, 4:1 (2001), pp. 1–16.

Bartlett, Robert. *'Gerald of Wales (c.1146–1220x23)'*, *ODNB* (Oxford: Oxford University Press, 2004), online edn. January 2008, http://www.oxforddnb.com/view/article/10769 (accessed 27th June 2020).

Bendall, Sarah. *'Speed, John (1551/2–1629)'*, *ODNB* (Oxford: Oxford University Press, 2004), online edn. January 2008, http://www.oxforddnb.com/view/article/26093 (accessed 3rd December 2013).

Beverley, Tessa. *'Blundeville, Thomas (1522?–1606?)'*, *ODNB* (Oxford: Oxford University Press, 2004), online edn. January 2008, http://www.oxforddnb.com/view/article/2718 (accessed 7th May 2021).

Bianco, Marcie. 'To Sodomize a Nation: *Edward II*, Ireland, and the Threat of Penetration', *EMLS*, 11 (1997), pp. 11.1–21.

Blundeville, Thomas. *A Briefe Description of the Universal Mappes and Cardes, and of Their Uses* (London: Roger Ward, 1589).

— *M. Blundevile His Exercises* (London: John Windet, 1594).

Boesky, Amy. *Founding Fictions: Utopias in Early Modern England* (Athens, Georgia: University of Georgia Press, 1996).

Borlik, Todd. *Ecocriticism and Early Modern English Literature: Green Pastures* (London: Routledge, 2010).

Boyle, Robert. *Some Considerations Touching the Vsefulnesse of Experimental Naturall Philosophy* (Oxford: Henry Hall, 1663).

— *The Christian Virtuoso Shewing That by Being Addicted to Experimental Philosophy* (London: Edward Jones, 1690).

— *Selected Philosophical Papers of Robert Boyle*, ed. M.A. Stewart (Manchester: Manchester University Press, 1979).

Brink, Jean. *Michael Drayton Revisited* (Boston: Twayne, 1990).

Brotton, Jerry. 'Mapping the Early Modern Nation: Cartography Along the English Margins', *Paragraph*, 19:2 (1996), pp. 139–155.

— 'Tragedy and Geography', *A Companion to Shakespeare's Works. Volume 1: The Tragedies*, eds. Richard Dutton and Jean E. Howard (Oxford: Blackwell Publishing Ltd., 2003), pp. 219–240.

— 'Printing the Map, Making a Difference: Mapping the Cape of Good Hope, 1488–1652', *Geography and Revolution*, eds. David N. Livingstone and Charles W. Withers (Chicago: University of Chicago Press, 2005), pp. 137–159.

— *A History of the World in Twelve Maps* (London: Penguin, 2012).

Bruce, Susan. ed. *Three Early Modern Utopias: Thomas More's 'Utopia', Francis Bacon's 'New Atlantis' and Henry Neville's 'The Isle of Pines'* (Oxford: Oxford University Press, 2008).

Bucholz, Robert, and Newton, Key. *Early Modern England 1485–1714: A Narrative History*, 2nd edition (Chichester: Wiley & Sons Ltd., 2009).

Buisseret, David. *The Mapmaker's Quest: Depicting New Worlds in Renaissance Europe* (Oxford: Oxford University Press, 2003).

Bundy, Murray W. '"Invention" and "Imagination" in the Renaissance', *The Journal of English and Germanic Philology*, 29:4 (1930), pp. 535–545.

Burnett, Mark Thornton. 'Tamburlaine: An Elizabethan Vagabond', *Studies in Philology*, 84:3 (1987), pp. 308–323.

Burton, Robert. *The Anatomy of Melancholy* (Oxford: Henry Cripps, 1621).

Camden, William. *Britannia siue Florentissimorum regnorum, Angliae, Scotiae, Hiberniae, et insularum adiacentium ex intima antiquitate chorographica descriptio'. Camden Britannia* (London: Ralph Newbery, 1586).

— *Britain, or A chorographicall description of the most flourishing kingdomes, England, Scotland, and Ireland, and the ilands adjoyning, out of the depth of antiquitie beautified vvith mappes of the severall shires of England*, trans. Philemon Holland (London: George Latham, 1637).

Canny, Nicholas. 'Reviewing a View of the Present State of Ireland', *Irish University Review*, 26:2 (1996), pp. 252–267.

— *Making Ireland British 1580–1650* (Oxford: Oxford University Press, 2001).

— 'The Origins of Empire: An Introduction', *The Oxford History of the British Empire, Volume 1: The Origins of Empire*, ed. Nicholas Canny (Oxford: Oxford University Press, 2001), pp. 1–33.

Carpenter, Nathaniel. *Geographie delineated forth in two bookes* (Oxford: John Lichfield, 1635).

Carroll, Clare. 'Spenser and the Irish Language: The Sons of Milesio in a View of the Present State of Ireland, The Faerie Queene, Book V and the Leabhar Gabhála', *Irish University Review*, 26:2 (1996), pp. 281–290.

Cavanagh, Sheila T. *Wanton Eyes and Chaste Desires: Female Sexuality in 'The Faerie Queene'* (Bloomington: Indiana University Press, 1994).

Certeau, Michel de. *The Writing of History* trans. Tom Conley (New York: Columbia University Press, 1988).

Chapple, Anne S. 'Robert Burton's Geography of Melancholy', *Studies in English Literature 1500–1900*, 33:1 (1993), pp. 99–130.

Charlton, Kenneth. *Education in Renaissance England* (London: Routledge, 1965).

Coate, Mary. *Cornwall in the Great Civil War and Interregnum, 1642–1660: A Social and Political Study* (Truro: Barton, 1963).

Cogswell, Thomas. 'The Path to Elizium "Lately Discovered": Drayton and the Early Stuart Court', *The Huntington Library Quarterly*, 54:3 (1991), pp. 207–233.

Coiro, Ann Baynes. *Robert Herrick's Hesperides and the Book Epigram Tradition* (Baltimore: Johns Hopkins University Press, 1988).

Conrad, Joseph. *'Geography and Some Explorers', Last Essays*, ed. Richard Curle (London & Toronto: J. M. Dent & Sons, 1926), pp. 10–17.

Corbett, Margery and Lightbown, Ronald. *The Comely Frontispiece: The Emblematic Title-Page in England 1550–1660* (London: Routledge, 1979).

Cormack, Lesley B. *Charting an Empire: Geography at the English Universities, 1580–1620* (Chicago: Chicago University Press, 1997).

— 'Britannia Rules the Waves?: Images of Empire in Elizabethan England', *EMLS*, 4.2:3 (1998), pp. 10.1–20.

— 'Maps as Educational Tools in the Renaissance', *The History of Cartography, Volume Three, Part 1: Cartography in the European Renaissance*, ed. David Woodward (Chicago: Chicago University Press, 2007), pp. 622–636.

Coughlan, Patricia. '"Cheap and Common Animals": The English Anatomy of Ireland in the Seventeenth Century', *Literature and the English Civil War*, ed. Thomas F. Healy and Jonathan Sawday (Cambridge: Cambridge University Press, 1990), pp. 205–223.

Covella, Francis Dolores. 'The Choral Nexus in Doctor Faustus', *Studies in English Literature, 1500–1900*, 26:2 (1986), pp. 201–215.

Cowper, William. *Three Heavenly Treatises, Concerning Christ* (London: Printed by T.S. for John Budge, 1612).

Cranford, James. *The Teares of Ireland* (London: John Rothwell, 1642).

Cuningham, William. *The Cosmographical Glasse Conteinyng the Pleasant Principles of Cosmographie, Geographie, Hydrographie, or Nauigation* (London: John Day, 1559).

Curtis, Mark. *Oxford and Cambridge in Transition: 1558–1642* (Oxford: Clarendon Press, 1959).

Cust, Lionel. 'Notes on the Collections Formed by Thomas Howard, Earl of Arundel and Surrey', *The Burlington Magazine for Connoisseurs*, 19:101 (1911), pp. 97–100.

Dailey, Jeff. 'Christian Underscoring in Tamburlaine the Great, Part II', *The Journal of Religion and Theatre*, 4:2 (1995), pp. 146–159.

Daniel, Samuel. *A Defence of Ryme* (London: R. Read, 1603).

Davis, J.C. *Utopia and the Ideal Society: A Study of English Utopian Writing, 1516–1700* (Cambridge: Cambridge University Press, 1983).

Das, Nandini *Renaissance Romance: The Transformation of English Prose Fiction, 1570–1620* (Surrey: Ashgate, 2011).

— 'Romance Re-Charted: The "Ground-Plots" of Sidney's *Arcadia*', *The Yearbook of English Studies*, 41:1 (2011), pp. 51–67.

Daston, Lorraine and Galison, Peter. 'The Image of Objectivity', *Representations*, 40 (1992), pp. 81–128.

Dean, Leonard F. 'Sir Francis Bacon's Theory of Civil History-Writing', *ELH*, 8:3 (1941), pp. 161–183.

Deats, Sara Munson. 'Myth and Metamorphosis in Marlowe's *Edward II*', *Texas Studies in Literature and Language*, 22:3 (1980), pp. 304–321.

Dee, John. *The Elements of Geometrie of the Most Auncient Philosopher Euclide of Megara* (London: John Day, 1570).

Derricke, John. *The Image of Irelande with a Discouerie of Woodkarne* (London: John Day, 1581).

Digges, Leonard and Thomas *A Geometrical Practical Treatize Named Pantometria* (London: Abell Jeffes, 1571).

Dillon, Michael. 'Governing through Contingency: the Security of Biopolitical Governance', *Political Geography*, 26:1 (2007), pp. 41–47.

Dollimore, Jonathan. *Radical Tragedy: Religion, Ideology, and Power in the Drama of Shakespeare and His Contemporaries* (North Carolina: Duke University Press, 2004).

Donne, John. *The Major Works*, ed. John Carey (Oxford: Oxford University Press, 1990).

Eastman, Nate. 'The Rumbling Belly Politic: Metaphorical Location and Metaphorical Government in *Coriolanus*', *EMLS*, 13:1 (2007), pp. 2.1–39.

Easton, Joy B. 'Leonard Digges', *Dictionary of Scientific Biography, Volume Four*, ed. C. C. Gillispie (New York: Charles Scribner's Sons, 1980), p. 97.

Eden, Richard. *A Briefe Collection and Compendious Extract of the Straunge and Memorable Things, Gathered Oute of the Cosmographye of Sebastian Munster* (London: Thomas Marshe, 1572).

— *History of Travayl* (London: Richarde Jugge, 1577).

Eerde, Katherine S. Van. *Wenceslaus Hollar: Delineator of His Time* (Virginia: Folger Books, 1970).

Eisenstein, Elizabeth. *The Printing Press as an Agent of Change* (Cambridge: Cambridge University Press, 1980).

Elyot, Thomas. *The Boke Named the Governour* (London: Thomas Berthelet, 1537).

Emberton, Wilfred. *The English Civil War Day By Day* (Stroud: Alan Sutton, 1995).

Emery, F.V. 'Irish Geography in the Seventeenth Century', *Irish Geography*, 3:5 (1958), pp. 263–276.

Empson, William. 'Donne the Space Man', *The Kenyon Review*, 19:3 (1957), pp. 337–399.

Erasmus, Desiderius. *The Correspondence of Erasmus: Letters 594–841 (1517–1518)* (Toronto: University of Toronto Press, 1979).

Ewell, Barbara C. 'Drayton's *Poly-Olbion*: England's Body Immortalized', *Studies in Philology*, 75:3 (1978), pp. 297–315.

Fennell, Barbara. '"Dodgy Dossiers"? Hearsay and the 1641 Depositions', *History Ireland*, 19:3 (2011), pp. 26–29.

Ferguson, Margaret W., Maureen, Quilligan and Vickers, Nancy J. *Rewriting the Renaissance: The Discourses of Sexual Difference in Early Modern Europe* (Chicago: University of Chicago Press, 1986).

Fisch, H. and Jones, H.W. 'Bacon's Influence on Sprat's *History of the Royal Society*', *Modern Language Quarterly*, 12 (1951), pp. 399–406.

Fleming, James Dougal. ed. *The Invention of Discovery, 1500–1700* (Surrey: Ashgate, 2011).

Fletcher, David. '*Saxton, Christopher (1542x4–1610–11)*', *ODNB* (Oxford: Oxford University Press, 2004), online edn. January 2008, http://www.oxforddnb.com/view/article/24760 (accessed 3rd May 2021).

Gailhard, Jean. *The Compleat Gentleman, or, Directions for the Education of Youth* (London: Thomas Newcomb, 1678).

Gernon, Luke. A Discourse on Ireland, anno. 1620 (1620), http://www.ucc.ie/celt/published/E620001/ (accessed 12th March 2013).

Giglioni, Guido. 'Francis Bacon', *The Oxford Handbook of British Philosophy in the Seventeenth Century*, ed. Peter R. Anstey (Oxford: Oxford University Press, 2013), pp. 41–72.

Gilbert, Humphrey. *Queene Elizabethes Achademy*, ed. F. J. Furnivall (London: Early English Text Society, 1869).

Gillies, John. *Shakespeare and the Geography of Difference* (Cambridge: Cambridge University Press, 1994).

— 'Marlowe, the Timur Myth and the Motives of Geography', *Playing the Globe: Genre and Geography in English Renaissance Drama*, eds. John Gillies and Virginia Mason Vaughan (New Jersey: Fairleigh Dickinson University Press, 1998), pp. 203–229.

Gillingham, John. 'Images of Ireland 1170–1600: The Origins of English Colonialism', *History Today*, 37 (1987), pp. 16–22.

Gordon, Andrew and Klein, Bernhard. eds. *Literature, Mapping and the Politics of Space in Early Modern England* (Cambridge: Cambridge University Press, 2001).

Gorton, Lisa. 'John Donne's Use of Space', *EMLS*, 4:2, Special Issue 3 (1998), pp. 9.1–27.

Grafton, Anthony and Jardine, Lisa. 'Studied for Action: How Gabriel Harvey Read His Livy', *Past and Present*, 129 (1990), pp. 30–78.

Grande, Troni Y. *Marlovian Tragedy: The Play of Dilation* (Lewisburg: Bucknell University Press, 1999).

Gray, Austin K. 'Some Observations on Christopher Marlowe, Government Agent', *PMLA*, 43:3 (1928), pp. 682–700.

Greenblatt, Stephen. *Renaissance Self-Fashioning* (Chicago: Chicago University Press, 1980).

Greengrass, Mark, Leslie, Michael and Raylor, Timothy. eds. *Samuel Hartlib and Universal Reformation: Studies in Intellectual Communication* (Cambridge: Cambridge University Press, 1994).

Grenfell, Joanne Woolway. 'Significant Spaces in Edmund Spenser's View of the Present State of Ireland', *EMLS*, 4:2 (1998), pp. 6.1–21.

Griffiths, Antony and Kesnerová, Gabriela. *Wenceslaus Hollar: Prints and Drawings* (London: British Museum Publications, 1983).

Gwyn, David. 'Richard Eden Cosmographer and Alchemist', *The Sixteenth Century Journal*, 15:1 (1984), pp. 13–34.

Hadfield, Andrew. 'Another Look at Serena and Irena', *Irish University Review*, 26:2 (1996), pp. 291–302.

—— *Edmund Spenser's Irish Experience: Wilde Fruit and Salvage Soyl* (Oxford: Oxford University Press, 1997).

—— 'Spenser, Drayton, and the Question of Britain', *The Review of English Studies*, 51:204 (2000), pp. 582–599.

—— '*Derricke, John (fl. 1578–1581)*', *ODNB* (Oxford: Oxford University Press, 2004), online edn. January 2008,http://www.oxforddnb.com/view/article/7537 (accessed 3rd March 2021).

—— '*Spenser, Edmund (1552?–1599)*', *ODNB*, (Oxford: Oxford University Press, 2004), online edn. January 2008,http://www.oxforddnb.com/view/article/26145 (accessed 2nd April 2021).

—— *Edmund Spenser: A Life* (Oxford: Oxford University Press, 2012).

Hakluyt, Richard. *The principal nauigations, voyages, traffiques and discoueries of the English Nation* (London: George Bishop, 1589).

Hall, Marie B. 'Science in the Early Royal Society', *The Emergence of Science in Western Europe*, ed. M. Crosland (London: Macmillan, 1975), pp. 57–78.

Hardin, Richard. *Michael Drayton and the Passing of Elizabethan England* (Lawrence: University Press of Kansas, 1973).

Harding, Robert J.D. '*Hollar, Wenceslaus (1607–1677)*', *ODNB* (Oxford: Oxford University Press, 2004), online edn. January 2008, http://www.oxforddnb.com/view/article/13549 (accessed 20th December 2013).

Harley, J.B. and Woodward, David. *The History of Cartography, Volume 1* (Chicago: University of Chicago Press, 1987).

—— *The New Nature of Maps: Essays in the History of Cartography*, ed. Paul Laxton with an introduction byJ. H. Andrews (Baltimore and London: Johns Hopkins University Press, 2001).

Hartley, Craig. 'The Young Hollar in Prague: A Group of New Attributions', *Print Quarterly*, 8:3 (1991), pp. 252–274.

Harvey, David. 'Cosmopolitanism and the Banality of Geographical Evils', *Public Culture*, 12:2 (2000), pp. 529–564.

Harvey, P.D.A. *Maps in Tudor England* (Chicago: Chicago University Press, 1993).

Haskell, Yasmin. ed. *Diseases of the Imagination and Imaginary Disease in the Early Modern Period* (Turnhout: Brepols, 2011).

Haynes, Alan. *Invisible Power: The Elizabethan Secret Services, 1570–1603* (New York: St. Martin's Press, 1992).

Hazlitt, William. 'Lectures on the Dramatic Literature in the Age of Elizabeth', *Marlowe: The Critical Heritage, 1588–1896*, ed. Millar McClure (London: Routledge, 1979), pp. 77–81.

Heath, Robert. *Paradoxical Assertions and Philosophical Problems Full of Delight and Recreation for All Ladies and Youthful Fancies* (London: Printed by R.W for Charles Webb, 1659).

Helgerson, Richard. 'The Land Speaks: Cartography, Chorography, and Subversion in Renaissance England', *Representations*, 16 (1986), pp. 50–85.

—— *Forms of Nationhood: The Elizabethan Writing of England* (Chicago: University of Chicago Press, 1994).

—— 'Introduction', *Early Modern Literary Studies*, 4.2:3 (1998), pp. 1.1–14.

Heninger, S.K. 'Tudor Literature of the Physical Sciences', *Huntington Library Quarterly*, 32 (1969), pp. 101–33.

— *The Cosmographical Glasse: Renaissance Diagrams of the Universe* (California: Huntington Library Press, 1977).

Herrick, Robert. *The Complete Poetry of Robert Herrick*, ed. Tom Cain and Ruth Connolly (Oxford: Oxford University Press, 2013).

Highley, Christopher. *Shakespeare, Spenser, and the Crisis in Ireland* (Cambridge: Cambridge University Press, 1997).

Higton, H.K. *'Hood, Thomas (bap. 1556, d. 1620)', ODNB* (Oxford: Oxford University Press, 2004), online edn. January 2008, http://www.oxforddnb.com/view/article/13680 (accessed 21st February 2021).

Hind, Arthur M. *Wenceslaus Hollar* (London: John Lane, 1922).

Hobbes, Thomas. *Leviathan*, ed. with an introduction and notes by J.C.A. Gaskin (Oxford: Oxford University Press 2008).

Hodgen, Margaret T. 'Sebastian Münster (1489–1552): A Sixteenth-Century Ethnographer', *Osiris*, 11 (1954), pp. 504–529.

Holland, Peter. 'Mapping Shakespeare's Britain', *Shakespeare's Histories and Counter-Histories*, eds. Dermot Cavanagh, Stuart Hampton-Rees and Stephen Longstaffe (Manchester: Manchester University Press, 2006), pp. 198–218.

— 'The Dramatic Forms of Journey in English Renaissance Drama', *Travel and Drama in Shakespeare's Time*, eds. Jean-Pierre Maquerlot and Michèle Willems (Cambridge: Cambridge University Press, 2006), pp. 160–178.

Hollis, Gavin. '"Give Me the Map There": *King Lear* and Cartographic Literacy in Early Modern England', *The Portolan*, 68 (2007), pp. 8–25.

Hood, Thomas. The Use of the Celestial Globe in Plano, *Set Foorth in Two Hemispheres* (London: John Windet, 1590).

— *The Use of Both the Globes Celestiall and Terrestriall* (London: Thomas Dawson, 1592).

Hopkins, Lisa. *Christopher Marlowe: A Literary Life* (London: Palgrave Macmillan, 2000).

— *Christopher Marlowe, Renaissance Dramatist* (Edinburgh: Edinburgh University Press, 2008).

Horden, John. *'Peacham, Henry (b. 1578, d. in or after 1644)', ODNB* (Oxford University Press, 2004), online edn. January 2008, http://www.oxforddnb.com/view/article/21667 (accessed 7th May 2021).

Householder, Michael. 'Eden's Translations: Women and Temptation in Early America', *Huntingdon Library Quarterly*, 70:1 (2007), pp. 11–36.

Hovey, Kenneth Alan. '"Mountaigny Saith Prettily": Bacon's French and the Essay', *PMLA*, 106:1 (1991), pp. 71–73.

Hunter, Michael. *Science and Society in Restoration England* (Cambridge: Cambridge University Press, 1981).

Hunter, Matthew. 'Hooke's Figurations: A Figural Drawing Attributed to Robert Hooke', *Notes and Records of the Royal Society*, 64:3 (2010), pp. 251–260.

Hutson, Lorna. 'Fortunate Travelers: Reading for the Plot in Sixteenth-Century England', *Representations*, 41 (1993), pp. 83–103.

Hyde, Ralph. *Gilded Scenes and Shining Prospects: Panoramic Views of British Towns*, 1575–1900 (Connecticut: Yale Center for British Art, 1985), p. 11.

Ivic, Christopher. *'Mapping British Identities: Speed's Theatre of The Empire of Great Britaine', British Identities and English Renaissance Literature*, eds. David J. Baker and Willy Maley (Cambridge: Cambridge University Press, 2002), pp. 135–156.

Jacob, Christian. 'Theoretical Aspects of the History of Cartography: Toward a Cultural History of Cartography', *Imago Mundi*, 48:1 (1996), pp. 191–198.

Kolodny, Annette. *The Lay of the Land: Metaphor as Experience and History in American Life and Letters* (North Carolina: University of North Carolina Press, 1975).

Jardine, Lisa. *Francis Bacon: Discovery and the Art of Discourse* (Cambridge: Cambridge University Press, 1974).

— *Ingenious Pursuits: Building the Scientific Revolution* (Norfolk: Doubleday, 1999).

— and Stewart, Allan. *Hostage to Fortune: The Troubled Life of Francis Bacon, 1561–1626* (London: Gollancz, 1998).

Johnson, Francis R. 'Marlowe's Astronomy and Renaissance Skepticism', *ELH*, 13:4 (1946), pp. 241–254.

Jones, Emrys. '"A World of Ground": Terrestrial Space in Marlowe's Tamburlaine Plays', *The Yearbook of English Studies*, 38:1–2 (2008), pp. 168–182.

Kelly, L.G. *'Hoby, Sir Thomas (1530–1566)', ODNB* (Oxford University Press, 2004), online edn. January 2008, http://www.oxforddnb.com/view/article/13414 (accessed 4th May 2021).

King, John N. 'Queen Elizabeth I: Representations of the Virgin Queen', *Renaissance Quarterly*, 43:1 (1990), pp. 30–74.

— 'Spenser's Religion', *The Cambridge Companion to Spenser*, ed. Andrew Hadfield (Cambridge: Cambridge University Press, 2001), pp. 217–236.

Kinney, Arthur F. *Shakespeare's Webs: Networks of Meaning in Renaissance Drama* (New York: Routledge, 2004).

Kirschbaum, Leo. 'Marlowe's *Faustus*: A Reconsideration', *Review of English Studies*, 19:75 (1943), pp. 225–241.

Kish, George. 'The Cosmographic Heart: Cordiform Maps of the 16th Century', *Imago Mundi*, 19 (1965), pp. 13–21.

Kitchen, Frank. 'John Norden (c. 1547–1625): Estate Surveyor, Topographer, County Mapmaker and Devotional Writer', *Imago Mudi*, 49 (1997), pp. 43–61.

— *'Norden, John (c.1547–1625)', ODNB* (Oxford University Press, 2004), online edn. January 2008, http://www.oxforddnb.com/view/article/20250 (accessed 12th August 2020).

Klein, Bernhard. 'Partial Views: Shakespeare and the Map of Ireland', *EMLS*, 4.2:3 (1998).

— *Maps and the Writing Space of Space in Early Modern England and Ireland* (New York: Palgrave Macmillan, 2001).

Knapp, James A. *Illustrating the Past in Early Modern England* (Surrey: Ashgate, 2003).

Kocher, Paul H. 'The Old Cosmos: A Study in Elizabethan Science and Religion', *The Huntington Library Quarterly*, 15:2 (1952), pp. 101–121.

Kratochvíl, Miloš V. *Hollar's Journey on the Rhine* (Prague: Artia, 1965).

Krogt, Peter van der. 'Commercial Cartography in the Netherlands with Particular Reference to Atlas Production (16th–18th Centuries)', *La Cartografia dels Paisos Baixos* (Barcelona: Institut Cartogràfic de Catalunya, 1994), pp. 71–140.

Kuriyama, Constance Brown. *Christopher Marlowe: A Renaissance Life* (New York: Cornell University Press, 2010).

Kusukawa, Sachiko. 'Bacon's Classification of Knowledge', *Cambridge Companion to Francis Bacon*, ed. Markku Peltonen (Cambridge: Cambridge University Press, 1996), pp. 47–74.

Labriola, Albert C. 'Painting and Poetry of the Cult of Elizabeth 1st: The Ditchley Portrait and Donne's "Elegie: Going to Bed"', *Studies in Philology*, 93:1 (1996), pp. 42–63.

Lambarde, William. *A Perambulation of Kent* (London: Ralph Newberie, 1576 [1570]).

Larson, Jenifer. *Greek Nymphs: Myths, Culture and Lore* (Oxford: Oxford University Press, 2001).

Law, Andrew Bonar. *The Printe Maps of Ireland to 1612* (New Jersey: Eagle Press, 1983).

Lee, Patricia Ann. 'Some English Academies: An Experiment in the Education of Renaissance Gentlemen', *History of Education Quarterly*, 10:3 (1970), pp. 273–286.

Lensing, Dennis M. 'Postmodernism at Sea: The Quest for Longitude in Thomas Pynchon's *Mason & Dixon* and Umberto Eco's *The Island of the Day Before*', *The Multiple Worlds of Pynchon's Mason & Dixon: Eighteenth-Century Contexts, Postmodern Observations*, ed. Elizabeth Jane Wall Hinds (New York: Camden House, 2005), pp. 125–146.

Levy, F.J. 'Henry Peacham and the Art of Drawing', *Journal of the Warburg and Courtauld Institutes*, 37 (1974), pp. 174–190.

Lievsay, John León. 'A Word about Barnaby Rich', *The Journal of English and Germanic Philology* (1956), pp. 381–392.

Livingstone, David N. 'Geography, Tradition and the Scientific Revolution: An Interpretative Essay', *Transactions of the Institute of British Geographers*, 15:3 (1990), pp. 359–373.

Lupton, Julia 'Mapping Mutability: Or, Spenser's Irish plot', *Representing Ireland: Literature and the Origins of Conflict 1534–1660* eds. Brendan Bradshaw, Andrew Hadfield and Willy Maley (Cambridge: Cambridge University Press, 1993), pp. 93–115.

Maley, Willy. 'Spenser and Ireland: A Select bibliography', *Spenser Studies: A Renaissance Poetry Annual*, 9 (1991), pp. 227–242.

— 'Spenser and Ireland: An Annotated Bibliography, 1986–96', *Irish University Review*, 26:2 (1996), pp. 342–353.

— *Salvaging Spenser: Colonialism, Culture and Identity* (London: Macmillan Press, 1997).

— *Nation, State and Empire in English Renaissance Literature: Shakespeare to Milton* (Hampshire: Palgrave Macmillan, 2003).

Mandel, Jerome. 'Dream and Imagination in Shakespeare', *Shakespeare Quarterly*, 24:1 (1973), pp. 61–68.

Mandelbrote, Giles. 'John Evelyn and His Books', *John Evelyn and His milieu*, eds. Frances Harris and Michael Hunter (London: British Library, 2003), pp. 71–90.

Margey, Annaleigh. 'Visualising the Plantation: Mapping the Changing Face of Ulster', *History Ireland*, 17:6 (2009), pp. 42–45.

— 'A Hitherto Unknown Sketch Map by Lord Burghley', *Imago Mundi*, 64:1 (2012), pp. 96–100.

Marlowe, Christopher. *Tamburlaine the Great: In Two Parts*, ed. Una Ellis-Fermor (London: Methuen, 1930).

— *Doctor Faustus and Other Plays*, eds. David Bevington and Eric Rasmussen (Oxford: Oxford University Press, 1995).

Martin, F.X. 'Gerald of Wales, Norman Reporter on Ireland', *Studies: An Irish Quarterly Review* (1969), pp. 279–292.

Martin, Julian. *Francis Bacon, the State, and the Reform of Natural Philosophy* (Cambridge: Cambridge University Press, 1992).

Matei-Chesnoiu, Monica. *Early Modern Drama and the Eastern European Elsewhere: Representations of Liminal Locality in Shakespeare and His Contemporaries* (New Jersey: Associated University Press, 2009).

Mayhew, Robert J. *Enlightenment Geography: The Political Languages of British Geography*, 1650–1850 (London: Palgrave Macmillan, 2000).

McAlindon, Tom. 'Classical Mythology and Christian Tradition in Marlowe's *Doctor Faustus*', *Publications of the Modern Language Association of America*, 81:3 (1966), pp. 214–223.

McCabe, Richard A. 'Edmund Spenser, Poet of Exile', *Proceedings of the British Academy*, 80 (1991), pp. 73–103.

McClintock, Anne. *Imperial Leather: Race, Gender and Sexuality in the Colonial Contest* (New York: Routledge, 1995).

McConnell, Anita. '*Billingsley, Sir Henry (d. 1606)*', *ODNB* (Oxford: Oxford University Press, 2004), online edn. January 2008, http://www.oxforddnb.com/view/article/2392 (accessed 3rd May 2021).

McDermott, James. '*Frobisher, Sir Martin (1535?–1594)*', *ODNB* (Oxford: Oxford University Press, 2004), online edn. January 2008, http://www.oxforddnb.com/view/article/10191 (accessed 23rd February, 2019).

McInnis, David. *Mind-Travelling and Voyage Drama in Early Modern England* (Basingstoke: Palgrave Macmillan, 2012).

McLane, Paul E. 'Was Spenser in Ireland in Early November 1579?', *Notes and Queries*, 204 (1959), pp. 99–101.

McLean, Antonia. *Humanism and the Rise of Science in Tudor England* (London: Heinemann, 1972).

McLean, Matthew Adam. *The Cosmographia of Sebastian Münster: Describing the World in the Reformation* (Surrey: Ashgate, 2007).

McLeod, Bruce. *The Geography of Empire in English Literature 1580–1745* (Cambridge: Cambridge University Press, 1999).

McLuhan, Marshall. *The Gutenberg Galaxy: The Making of Typographic Man* (Toronto: University of Toronto Press, 1962).

McRae, Andrew and Schwyzer, Philip, eds. *Poly-Olbion: New Perspectives* (London: Boydell & Brewer, 2020).

Mendyk, S. 'Gerard Boate and Irelands Naturall History', *The Journal of the Royal Society of Antiquaries of Ireland*, 115 (1985), pp. 5–12.

Mendyk, Stan. 'Early British Chorography', Early British Chorography', *The Sixteenth Century Journal*, 17:4 (1986), pp. 459–481.

Mentz, Steve. *Romance for Sale in Early Modern England: The Rise of Prose Fiction* (Surrey: Ashgate, 2006).

Middleton, Thomas. *Collected Works of Thomas Middleton*, eds. Gary Taylor and John Lavagnino (Oxford: Oxford University Press, 2007).

Miller, Paul W. 'A Function of Myth in Marlowe's "Hero and Leander"', *Studies in Philology*, 50:2 (1953), pp. 158–167.

Milton, John. *Of Education* (London: 1644).

— *Paradise Lost*, ed. David Scott Kastan (Indiana: Hackett Publishing Ltd., 2005).

— *Milton's Selected Poetry and Prose*, ed. Jason P. Rosenblatt (London: W. W. Norton and Company, 2011).

Montaigne, Michel de. *The Complete Essays*, trans. M.A. Screech (London: Penguin Classics, 2003).

Moran, Michael G. *Moran, Inventing Virginia: Sir Walter Raleigh and the Rhetoric of Colonization, 1584–1590* (New York: Peter Lang Publishing, 2007).

Moroney, Maryclaire. 'Apocalypse, Ethnography, and Empire in John Derricke's *Image of Irelande* (1581) and Spenser's *View of the Present State of Ireland* (1596)', *English Literary Renaissance*, 29:2 (1999), pp. 355–374.

Morrissey, John. 'Foucault and the Colonial Subject: Emergent Forms of Colonial Governmentality in Early Modern Ireland', *At the Anvil: Essays in Honour of William J. Smyth*, eds. Patrick Duffy and William Nolan (Dublin: Geography Publications, 2012), pp. 135–150.

Mukherjee, Nilinjana. 'A desideratum More sublime': Imperialism's Expansive Vision and Lambton's Trigonometrical Survey of India', *Postcolonial Studies*, 14:4 (2011), pp. 429–447.

— *Spatial Imaginings in the Age of Colonial Cartographic Reason: Maps, Landscapes, Travelogues in Britain and India* (London: Taylor and Francis, 2020).

Murphy, Andrew. *But the Irish Sea Betwixt Us: Ireland, Colonialism, and Renaissance Literature* (Kentucky: University of Kentucky Press, 1999).

Murray, Penelope. 'Editor's Introduction', *Imagination: A Study in the History of Ideas*, ed. Barry Cocking (Routledge: London, 1990), pp. vi–xvi.

Myers, J.P., ed. *Elizabethan Ireland: A Selection of Writings by Elizabethan Writers on Ireland* (Connecticut: Archon, 1983).

Neill, Michael. '"Mulattos," "Blacks," and "Indian Moors": *Othello* and Early Modern Constructions of Human Difference', *Shakespeare Quarterly*, 49:4 (1998), pp. 361–374.

Netzloff, Mark. 'Forgetting the Ulster Plantation: John Speed's *The Theatre of the Empire of Great Britain* (1611) and the Colonial Archive', *Journal of Medieval and Early Modern Studies*, 31:2 (2001), pp. 313–348.

Newmark, Paige. '"She Is Spherical, Like a Globe": Mapping the Theatre, Mapping the Body', *Shakespeare in Southern Africa*, 16 (2004), pp. 15–28.

Nicholl, Charles. *'Marlowe, Christopher (bap. 1564, d. 1593)'*, *ODNB* (Oxford: Oxford University Press, 2004), online edn. January 2008, http://www.oxforddnb.com/view/article/18079 (accessed 23rd March 2021).

Norden, John. *Nordens preparatiue to His Speculum Britanniae* (London: Daniel Browne and James Woodman, 1596).

— *The Surveyor's Dialogue (1618): A Critical Edition*, ed. with an introduction by Mark Netzloff (Surrey: Ashgate, 2010).

Nuti, Lucia. 'Mapping Places: Chorography and Vision in the Renaissance', *Mappings*, ed. Denis Cosgrove (London: Reaktion Books, 2000), pp. 90–108.

Oates, J.C.T. *Cambridge University Library: A History* (Cambridge: Cambridge University Press, 1986).

O'Loughlin, Thomas. 'An Early Thirteenth-Century Map in Dublin: A Window into the World of Giraldus Cambrensis', *Imago Mundi*, 51 (1999), pp. 24–39.

O'Neill, Stephen. *Staging Ireland: Representations in Shakespeare and Renaissance Drama* (Dublin: Four Courts Press, 2007), pp. 90–103.

Ong, Walter J. 'Spenser's *View* and the Tradition of the "Wild" Irish', *Modern Language Quarterly*, 3:4 (1942), pp. 561–571.

— *Ramus: Method and the Decay of Dialogue* (Massachusetts: Harvard University Press, 1958).

Orgel, Stephen. 'Tobacco and Boys: How Queer Was Marlowe?', *GLQ: A Journal of Lesbian and Gay Studies*, 6:4 (2000), pp. 555–576.

Ortelius, Abraham. *Theatrum Orbis Terrarum* (Antwerp: Gilles Coppens de Diest, 1570).

Osgood, C.G. *Concordance to the Poems of Edmund Spenser* (Washington: Carnegie Institution of Washington, 1969).

Palmer, Patricia. *Language and Conquest in Early Modern Ireland: English Renaissance Literature and Elizabethan Imperial Expansion* (Cambridge: Cambridge University Press, 2001).

— *The Severed Head and the Grafted Tongue: Literature, Translation and Violence in Early Modern Ireland* (Oxford: Oxford University Press, 2014).

Park, Katherine. *'The Imagination in Renaissance Psychology'* (unpublished M. Phil. Thesis, University of London, 1974).

Parry, Graham. *'Ware, Sir James (1594–1666)'*, *ODNB* (Oxford: Oxford University Press, 2004), online edn. January 2008,http://www.oxforddnb.com/view/article/28729 (accessed 25th February 2020).

Peacham, Henry. *The Compleat Gentleman* (London: John Legat, 1622).

Pelletier, Monique. 'Representations of Territory by Painters, Engineers, and Land Surveyors in France during the Renaissance', *The History of Cartography, Volume Three, Part 1: Cartography in the European Renaissance*, ed. David Woodward (Chicago: Chicago University Press, 2007), pp. 1522–1537.

Peltonen, Markku. ed. *Cambridge Companion to Francis Bacon* (Cambridge: Cambridge University Press, 1996).

— *'Bacon, Francis, Viscount St Alban (1561–1626)'*, *ODNB* (Oxford: Oxford University Press, 2004), online edn. January 2008,http://www.oxforddnb.com/view/article/990 (accessed 11th May 2021).

Pemble, William. *A Briefe Introduction to Geography* (Oxford: John Lichfield, 1630).

Pennington, Richard. *A Descriptive Catalogue of the Etched Works of Wenceslaus Hollar 1607–1677* (Cambridge: Cambridge University Press, 1982).

Pérez-Ramos, Antonio. *Francis Bacon's Idea of Science and the Maker's Knowledge Tradition* (Oxford: Clarendon Press, 1991).

Plutarch. *The philosophie, commonlie Called, the morals vvritten by the Learned Philosopher Plutarch of Chaeronea*, trans. Philemon Holland (London: Arnold Hatfield, 1603).

Poole, William. 'Early Modern Eclecticism', *Critical Quarterly*, 52:4 (2010), pp. 12–22.

Powell, J.R. *The Navy in the English Civil War* (Hamden: Archon Books, 1962).

Prescott, Anne Lake. *'Drayton, Michael (1563–1631)'*, *ODNB* (Oxford: Oxford University Press, 2004), online edn. January 2008, http://www.oxforddnb.com/view/article/8042 (accessed 19th June 201).

Raleigh, Walter. *The History of the World* (London: William Stansby, 1617).

— *A Discovery of Guiana, The English Renaissance: An Anthology of Sources and Documents*, ed. Kate Aughterson (London: Routledge, 2002).

Ramus, Petrus. *The Way to Geometry Being Necessary and Usefull, for Astronomers*, trans. William Bedwell (London: Thomas Cotes, 1636).

Rapple, Rory. *'Gilbert, Sir Humphrey (1537–1583)'*, *ODNB* (Oxford: Oxford University Press, 2004), online edn. January 2008, http://www.oxforddnb.com/view/article/10690 (accessed 7th May 2021).

— *Martial Power and Elizabethan Political Culture: Military Men in England and Ireland, 1558–1594* (Cambridge: Cambridge University Press, 2009).

Rawley, William. *Resuscitatio, or, Bringing into publick light severall pieces of the Works, Civil, Historical, Philosophical, & Theological, Hitherto Sleeping, of the Right Honourable Francis Bacon, Baron of Verulam, Viscount Saint Alban* (London: Sarah Griffin, 1657).

Ribner, Irving. *'Marlowe's "Tragicke Glasse"', Essays on Shakespeare and Elizabethan Drama in Honor of Hardin Craig*, ed. Richard Hosley (Columbia: University of Missouri Press, 1962), pp. 91–114.

Richeson, A.W. *English Land Measuring to 1800: Instruments and Practices* (Massachusetts: Society for the History of Technology and MIT Press, 1966).

Riggs, David. *The World of Christopher Marlowe* (New York: Henry Holt Publishing, 2004).

Riss, Arthur. 'The Belly Politic: Coriolanus and the Revolt of Language', *ELH*, 59:1 (1992), pp. 53–75.

Rodgers, Thomas *The Anatomie of the Minde* (London: John Charlewood, 1576).

Rossi, Paolo. 'Bacon's Idea of science', *Cambridge Companion to Francis Bacon*, ed. Markku Peltonen (Cambridge: Cambridge University Press, 1996), pp. 25–46.

Rowse, A.L. *The Expansion of Elizabethan England* (Wisconsin: University of Wisconsin Press, 2003).

Rutter, Tom. *The Cambridge Introduction to Christopher Marlowe* (Cambridge: Cambridge University Press, 2012).

Said, Edward W. *Culture and Imperialism* (London: Vintage, 1994).

Salzman, Paul. 'Essays', *The Oxford Handbook of English Prose 1500–1640*, ed. Andrew Hadfield (Oxford: Oxford University Press, 2013), pp. 468–483.

Sanders, Julie. *The Cultural Geography of Early Modern Drama, 1620–1650* (Cambridge: Cambridge University Press, 2011).

Sanford, Rhonda Lemke. *Maps and Memory in Early Modern England: A Sense of Place* (London: Palgrave Macmillan, 2002).

Sargent, Rose-Mary. *'Bacon as an Advocate for Co-Operative Research', Cambridge Companion to Francis Bacon*, ed. Markku Peltonen (Cambridge: Cambridge University Press, 1996), pp. 46–171.

Sawday, Jonathan *The Body Emblazoned: Dissection and the Human Body in Renaissance Culture* (London: Routledge, 1995).

— and Rhodes, Neil, eds. *The Renaissance Computer: Knowledge Technology in the First Age of Print* (London: Routledge, 2002).

Seaton, Ethel. 'Marlowe's Map', *Essays and Studies*, 10 (1924), pp. 13–35.

— 'Fresh Sources for Marlowe', *The Review of English Studies*, 5:20 (1929), pp. 385–401.

Semler, L.E. 'Breaking the Ice to Invention: Henry Peacham's *The Art of Drawing* (1606)', *The Sixteenth Century Journal*, 35:3 (2004), pp. 735–750.

Seymour, M.C. *'Bartholomaeus Anglicus (b. before 1203, d. 1272)', ODNB* (Oxford: Oxford University Press, 2004), online edn. January 2008, http://www.oxforddnb.com/view/article/10791 (accessed 9th June 2021).

Shakespeare, William. *The Oxford Shakespeare: The Complete Works*, eds. John Jowett, William Montgomery, Gary Taylor and Stanley Wells, 2nd edition (Oxford: Oxford University Press, 2005).

Shalev, Zur and Burnett, Charles. *Ptolemy's 'Geography' in the Renaissance* (London: Warburg Institute Colloquia, 2011).

Shannon, William and Winstanley, Michael. 'Lord Burghley's Map of Lancashire Revisited, c. 1576–1590', *Imago Mundi*, 59:1 (2007), pp. 24–42.

Shapiro, Barbara J. *A Culture of Fact: England, 1550–1720* (New York: Cornell University Press, 2003).

Shaw, Renata. 'Broadsides of the Thirty Years' War', *The Quarterly Journal of the Library of Congress*, 32:1 (1975), pp. 2–24.

Sidney, Philip. *The Major Works*, ed. with an introduction and notes by Katherine Duncan-Jones (Oxford: Oxford University Press, 2008).

Silberman, Lauren. *Transforming Desire: Erotic Knowledge in Books III and IV of 'The Faerie Queene'* (Berkeley: California University Press, 1995).

Skelton, R.A. 'Hakluyt's Maps', *The Hakluyt Handbook: Volume 1*, ed. D.B. Quinn (Hakluyt Society, London: 1974), pp. 48–73.

Smith, Catherine Delano. 'Map Ownership in Sixteenth-Century Cambridge: The Evidence of Probate Inventories', *Imago Mundi*, 47 (1995), pp. 67–93.

Smith, Jonathan M. 'State Formation, Geography, and a Gentleman's Education', *Geographical Review*, 86:1 (1996), pp. 91–100.

Smith, Donald K. *The Cartographic Imagination in Early Modern England: Re-Writing the World in Marlowe, Spenser, Raleigh and Marvell* (Aldershot: Ashgate, 2008).

Smyth, William J. *Map-Making, Landscapes and Memory: A Geography of Colonial and Early Modern Ireland c. 1530–1750* (Indiana: University of Notre Dame Press, 2004).

Snider, Alvin. 'Bacon, *Legitimation* and the *"Origin of Restoration Science"'*, *The Eighteenth Century*, 32:2 (1991), pp. 119–138.

Sofer, Andrew J. 'How to Do Things with Demons: Conjuring Performatives in *Doctor Faustus*', *Theatre Journal*, 61:1 (2009), pp. 1–21.

Somogyi, Nick de. 'Marlowe's Maps of War', *Christopher Marlowe and English Renaissance Culture*, eds. Darryll Grantley and Peter Roberts (London: Scolar Press, 1996), pp. 96–109.

Speed, John. *A Description of the Civill Warres of England* (T. East, 1601).

— *The Theatre of the Empire of Great Britaine* (London: William Hall, 1611).

Speed, Jeffrey John. *Tudor Townscapes: The Town Plans from John Speed's 'Theatre of the Empire of Great Britaine 1610'* (Buckingham: Map Collector Publications Ltd, 2000).

Spenser, Edmund. *Three Proper And Familiar Letters* (London: H. Bynneman, 1580).

— *A View of the State of Ireland: From the First Printed Edition (1633)*, eds. Andrew Hadfield and Willy Maley (Oxford: Blackwell Publishing, 1997).

Spiller, Elizabeth. *Science, Reading and Renaissance Literature: The Art of Making Knowledge, 1580–1670* (Cambridge: Cambridge University Press, 2004).

Spinrad, Phoebe S. 'The Dilettante's Lie in *Doctor Faustus*', *Texas Studies in Literature and Language*, 24:3 (1982), pp. 243–254.

Sprat, Thomas. *The History of the Royal Society of London* (London: T. R. for J. Martyn., 1667).

Springell, Francis C. *Connoisseur and Diplomat: the Earl of Arundel's Embassy to Germany in 1636, with a Transcription of the Diaries of William Crowne* (London: Maggs Bros., 1963).

Stanihurst, Richard. *Great Deeds in Ireland: Richard Stanihurst's De Rebus in Hibernia Gestis*, eds. John Barry and Hiram Morgan (Cork: Cork University Press, 2013).

Starkey, David. 'The Court: Castiglione's Ideal and Tudor Reality; Being a Discussion of Sir Thomas Wyatt's Satire Addressed to Sir Francis Bryan', *Journal of the Warburg and Courtauld Institutes*, 45 (1982), pp. 232–239.

Stone, Lawrence. 'The Educational Revolution in England, 1560–1640', *Past & Present*, 28:1 (1964), pp. 41–80.

Strandsbjerg, Jeppe. 'The Cartographic Assemblage of the Globe' (Chicago: Working paper presented at the International Studies Association convention, 2007).

Sugg, Richard. *Murder After Death: Literature and Anatomy in Early Modern England* (Ithaca and London: Cornell University Press, 2007).

Sullivan, Garrett A. *The Drama of Landscape: Land, Property, and Social Relations on the Early Modern Stage* (California: Stanford University Press, 1998).

— 'Geography and Identity in Marlowe', *The Cambridge Companion to Christopher Marlowe*, ed. Patrick Cheney (Cambridge: Cambridge University Press, 2004), pp. 231–244.

Sumption, Jonathan. *Age of Pilgrimage: The Medieval Journey to God* (New Jersey: Paulist Press, 2003).

Taylor, E.G.R. *Tudor Geography: 1485–1583* (Oxford: Clarendon Press, 1930).

— *Late Tudor and Early Stuart Geography* (Oxford: Clarendon Press, 1934).

Thackeray, Anne. *Caterpillars and Cathedrals: The Art of Wenceslaus Hollar* (Toronto: Thomas Fisher Rare Book Library, 2010).

Thomas, Vivien and Tydeman, William. *Christopher Marlowe: The Plays and Their Sources* (London: Routledge, 1994).

Thrower, Norman J. *Maps and Civilization: Cartography in Culture and Society*, 3rd edition (Chicago: University of Chicago Press, 2008).

Tindall, Gillian. *The Man Who Drew London: Wenceslaus Hollar in Reality and Imagination* (London: Vintage, 2002).

Traub, Valerie. 'The Nature of Norms in Early Modern England: Anatomy, Cartography, *King Lear*', *South Central Review*, 26:1 and 2 (2009), pp. 42–81.

Turnbull, David. 'Cartography and Science in Early Modern Europe: Mapping the Construction of Knowledge Spaces', *Imago Mundi*, 48 (1996), pp. 5–24.

Turner, Simon. 'Hollar in Holland: Drawings from the Artist's Visit to the Dutch Republic in 1634', *Master Drawings*, 48 (2010), pp. 73–104.

— 'Hollar's Prospects and Maps of London', *Printed Images in Early Modern Britain: Essays in Interpretation*, ed. Michael Hunter (Surrey: Ashgate, 2010).

Tyacke, Sarah and Huddy, John. *Christopher Saxton and Tudor Map-Making* (London: British Library Reference Division, 1980).

Tyndale, William. *The whole workes of W. Tyndall, Iohn Frith, and Doct. Barnes* (London: John Daye, 1573).

Urry, William. *Christopher Marlowe and Canterbury* (London: Faber & Faber, 1988).

Urzidil, Johannes. *Hollar: A Czech Émigré in England* (London: Czechoslovak, 1942).

Veltman, Kim. *Military Surveying and Topography: The Practical Dimension of Renaissance Linear Perspective* (Coimbra: UC Biblioteca Geral, 1979).

Vernon, E.C. *'Cranford, James (1602/3–1657)'*, *ODNB* (Oxford University Press, 2004), online edn. January 2008, http://www.oxforddnb.com/view/article/6610 (accessed 12th August 2020).

Vertue, George. *A Description of the Works of Wenceslaus Hollar* (London: William Bathoe, 1745).

Vigo, Giovanni da. *The most excelent worckes of chirurgery*, trans. Bartholomew Traheron (London: Edward Whyte, 1543).

Voekel, Swen '"Upon the Suddaine View": State, Civil Society and Surveillance in Early Modern England', *EMLS*, 4.2: 2.1-27 (1998).

Voltaire, 'On Chancellor Bacon', *Philosophical Letters: Letters Concerning the English Nation, trans. Ernest Dilworth* (New York: Dover, 2003), pp. 46–51.

Waite, Gary K. *Heresy, Magic, and Witchcraft in Early Modern Europe* (New York: Palgrave Macmillan, 2003).

Waith, Eugene M. *The Herculean Hero in Marlowe, Chapman, Shakespeare and Dryden* (New York: Columbia University Press, 1962).

Walker, Julia M. 'The Visual Paradigm of "The Good-Morrow": Donne's Cosmographical Glasse', *The Review of English Studies*, 37:145 (1986), pp. 61–65.

Wallis, Helen '"Opera Mundi": Emery Molyneux, Jodocus Hondius and the First English Globes', *Theatrum Orbis Librorum: Liber Amicorum Presented to Nico Israel on the Occasion of His Seventieth Birthday*, eds. Ton Croiset van Uchelen, T.K. Croiset Van Uchelen, K. Van Der Horst, Gunter Schilder and Nico Israel (Utrecht: Hes & De Graff Publishing, 1989), pp. 94–104.

Warnicke, Retha M. '*Nowell, Laurence (1530–c.1570)*', *ODNB* (Oxford: Oxford University Press, 2004), online edn. January 2008, http://www.oxforddnb.com/view/article/69731 (accessed 27th June 2021).

Watson, Ruth. 'Cordiform Maps since the Sixteenth Century: The Legacy of Nineteenth Century Classificatory Systems', *Imago Mundi*, 60:2 (2008), pp. 182–194.

Weinberger, Jerry. *Science, Faith, and Politics: Francis Bacon and the Utopian Roots of the Modern Age* (Ithaca and London: Cornell University Press, 1988).

Weiss, Benjamin. 'The *Geography* in Print: 1475: 1530', *Ptolemy's 'Geography' in the Renaissance*, eds. Zur Shalev and Charles Burnett (London: Warburg Institute Colloquia 2011), pp. 91–120.

Wessman, Christopher. 'Marlowe's *Edward II* as "Actaeonesque History"', *Connotations*, 9:1 (1999), pp. 1–33.

Wiesner, Merry E. *Women and Gender in Early Modern Europe* (Cambridge: Cambridge University Press, 2000).

Wiggins, Martin. ed. *British Drama, 1533–1642: A Catalogue* (Oxford: Oxford University Press, 2012).

Williams, R. Grant. 'Disfiguring the Body of Knowledge: Anatomical Discourse and Robert Burton's *The Anatomy of Melancholy*', *English Literary History*, 68:3 (2001), pp. 593–613.

Wintle, Michael. 'Renaissance Maps and the Construction of the Idea of Europe', *Journal of Historical Geography*, 25:2 (1999), pp. 137–165.

Withers, Charles and Mayhew, Robert. 'Rethinking "Disciplinary" History: Geography in British Universities, c. 1580–1887', *Transactions of the Institute of British Geographers*, 27 (2002), pp. 11–29.

Wood, Denis. *Rethinking the Power of Maps* (New York: Guilford Press, 2012).

Woods, Gregory. 'Body, Costume, and Desire in Christopher Marlowe', *Journal of Homosexuality*, 23:1–2 (1992), pp. 69–84.

Woodward, David. '*Cartography and the Renaissance: Continuity and Change*', *The History of Cartography, Volume Three, Part 1: Cartography in the European Renaissance*, ed. David Woodward (Chicago: Chicago University Press, 2007), pp. 3–24.

— ed. *The History of Cartography. Volume 3: Cartography in the European Renaissance* (Chicago: Chicago University Press, 2007).

Zambelli, Paola. *White Magic, Black Magic in the European Renaissance* (Boston: Brill, 2007).

Zeitlin, Jacob. 'The Development of Bacon's *Essays*: With Special Reference to the Question of Montaigne's Influence upon Them', *The Journal of English and Germanic Philology*, 27:4 (1928), pp. 496–519.

Zika, Charles. *Exorcising Our Demons: Magic, Witchcraft, and Visual Culture in Early Modern Europe* (Boston: Brill, 2003).

Index

For Product Safety Concerns and Information please contact our EU
representative GPSR@taylorandfrancis.com
Taylor & Francis Verlag GmbH, Kaufingerstraße 24, 80331 München, Germany

www.ingramcontent.com/pod-product-compliance
Lightning Source LLC
Chambersburg PA
CBHW060250220326
41598CB00027B/4054

9 781032 060262